FLIGHT IDENTIFICATION OF EUROPEAN RAPTORS

When on the wing this species (Honey Buzzard)
may be easily distinguished from the common
buzzard by its hawk-like appearance, small head,
wings not so blunt, and longer tail.
GILBERT WHITE *The Natural History of Selborne*
Letter XLIII to Thomas Pennant Esquire

Flight Identification of European Raptors

R. F. PORTER, IAN WILLIS
STEEN CHRISTENSEN AND
BENT PORS NIELSEN

T. & A. D. POYSER
London

ISBN 0 85661 027 5

First published by T. & A. D. Poyser Limited
Town Head House, Calton,
Waterhouses, Staffordshire, England

First edition 1974
Second edition 1976
Second edition reprinted (with some revised plates) 1978
Third edition 1981
Third edition reprinted 1986
Third edition reprinted 1992 by T. & A. D. Poyser Ltd
24–28 Oval Road, London, NW1 7DX

United States Edition published by
Academic Press Inc., San Diego, CA 92101

This book has evolved from material which first appeared
as a series of papers by the authors in the journal *British Birds*,
Vols 64–66, 1971–73

Text set in 9/10 pt Linotron 202 Plantin, printed and bound
in Great Britain at The Bath Press, Avon.

Contents

List of Plates

Foreword

Field identification of birds has advanced by leaps and bounds during the last 35 years, moving during that time from the often tentative guesswork of a previous generation to the analytical approach of the present day. It all really began with the sections on 'Field-characters and General Habits' by the late B. W. Tucker in *The Handbook of British Birds* (1938–41). Nowadays there are at least three major European field guides on the market, whereas prewar observers had largely to grope their own way until *The Handbook* was published, and then had that alone until *A Field Guide to Birds in Britain and Europe* by Roger Peterson, Guy Mountfort and P. A. D. Hollom first appeared in 1954. Yet it is a tribute to Tucker's whole approach to field identification that most of the field guides and reference books to this day still draw heavily on his clear and succinct summaries based on years of critical fieldwork and international correspondence.

But Tucker would have been the first to admit that, if any one group posed limitations on his methods, it was the diurnal birds of prey. His descriptions tended to concentrate on plumage colours, and the modern field guides have been inclined to do likewise. When, however, in the late 1940's and early 1950's attention began to turn increasingly to bird of prey identification, as more observers travelled Europe and Africa, it began to be realised how greatly some raptors, especially the eagles and buzzards, vary in plumage and how much their shapes and wing positions change with the mode of flight of the moment, which in turn may depend largely upon wind and weather.

In Britain, where the sole eagle is the Golden, and where the Buzzard has to be distinguished only from the rare Rough-legged and Honey Buzzards, these difficulties may not matter too much. But in Europe there are four vultures, nine eagles, five buzzards (separating the Steppe Buzzard which, though only a subspecies, is often confused with the Long-legged Buzzard), three round winged hawks, three kites (including Black-winged), four harriers, the Osprey and ten falcons—in all, 38 species and a difficult subspecies.

Among British observers, it was P. J. Hayman who took the lead in identifying European birds of prey in flight, and for many years observers going abroad were supplied by him with beautifully executed sketches based on his own experiences in the Middle East and Europe. Richard Porter, Ian Willis and I were among those who found themselves carrying Hayman's sketches during expeditions to various parts of the Continent. Meanwhile, a significant paper was published in 1968 on the identification of *Aquila* eagles, with an English summary and English captions to twelve pages of drawings, by Steen Christensen, Bent Pors Nielsen, N. H. Christensen and L. H. Sorensen (*Dansk Orn. Foren. Tidsskr.*, 62: 68–94) and followed in 1970 by another on the buzzards and remaining eagles by the first two of those authors (*Dansk Orn. Foren. Tidsskr.*, 64: 1–44). Also in 1970 L. H. Sorensen began a Belgian series with the buzzards (*Aves*, 7: 29–43) and in 1971 Lars Svensson made an extremely useful contribution to the identification of Pallid and Montagu's Harriers (*Vår Fågelvärld*, 30: 106–122). This growing interest in raptor identification was further reflected in the publication of *Birds of Prey in the Field* by Roger Harkness and Colin Murdoch (1971). On the global plane, too, *Birds of Prey of the World* by Mary Louise Grossman and John Hamlet (1965) and *Eagles, Hawks and Falcons of the World* by Leslie Brown and Dean Amadon (1968) both included many flight sketches.

I believe, however, that all these earlier publications have had their limitations so fas as flight identification is concerned, and the editors of the monthly journal *British Birds* were very pleased when, in 1971, they were able to persuade two of the Danish experts in this field, Steen Christensen and Bent Pors Nielsen, and two British experts, Richard Porter and Ian Willis, the latter a most accomplished artist, to pool their knowledge in producing a series of eight articles covering the identification of all the European raptors. These eight articles were published over the three years 1971–73 and have now been updated and brought together in this book. Birds of prey have long been a special interest of mine and, as then executive editor of *British Birds*, I had the privilege of working with these four authors in seeing the articles through the press.

It will be a long time before the last word on bird of prey identification is written, but I regard the information in this book as easily the most helpful yet in sorting out raptors in flight and I believe that Mr Willis's sketches are at once the most beautifully drawn and the most accurate of the flight illustrations so far published of this important and difficult group. I am sure that most other observers will agree. Perhaps, dare I hope, it will lead in time to comprehensive guides of a similar high standard to the flight identification of raptors in Africa and the rest of the Palearctic, even worldwide.

JAMES FERGUSON-LEES

Introduction

The identification of birds of prey in flight will always be a problem. No one can ever feel confident when confronted with species from a group showing such diversity of plumage, whose silhouettes vary in subtle ways in different circumstances, and for which the challenge of identification is so often made at considerable range.

We four, the authors of this book, have been studying flight identification of raptors for at least nine years—mostly in Europe, but also in Asia and Africa. A chance meeting in autumn 1968 between B. P. N. and R. F. P. at the now legendary Camlica Hills of the Turkish Bosphorus brought the Danish and English teams together and it was agreed to co-operate in a series of papers covering all the European raptors, and Ian Willis agreed to illustrate them. These papers were eventually published in the journal *British Birds* and now appear, with amendments, for the first time as a comprehensive collection. In addition the book ends with a section summarising the legal status of birds of prey in Europe.

For convenience of reference we have re-arranged the material on the species into the following seven sections:

1: THE BUZZARDS AND HONEY BUZZARD

2: THE EAGLES AND OSPREY

3: THE HARRIERS AND KITES

4: THE VULTURES

5: THE LARGE FALCONS

6: THE SMALL FALCONS AND BLACK-WINGED KITE

7: THE ACCIPITERS

The groups are designed to bring together species which have somewhat similar field characters and between which confusion can often arise. It is not our intention to dwell on the details of generic characters: we respectfully suggest that anyone who does not know the differences between, say, buzzards and eagles should first look at more basic identification guides.

Before considering the first group in detail, we must emphasise again that birds of prey are difficult to identify in flight (and even more so when perched). The only way to achieve competence is to get to know them in the field and to learn by mistakes. The novice should first familiarise himself with plumage characters, even though these are highly variable in some species. Only when he is competent in this respect can he hope to start identifying them on shape and structure, and thus at greater distances. Even then he must beware that shapes and wing positions can be misleading and highly variable, often depending on weather conditions. Moult is also a problem. No one should ever expect to identify every bird of prey. Anybody who travels widely in Europe and identifies 70% of all the raptors that he sees is doing extremely well—trying to be too ambitious will lead only to mistakes and to inaccurate documentation that may take years to rectify.

We would like to thank all those people who provided us with information and constructive criticism whilst the series was being published in *British Birds*, particularly those who responded to our requests for comments: C. A. Bauer, J. Bruun,

J. Cantelo, Arthur Christiansen, R. H. Dennis, L. Eccles, P. R. Flint, P. A. Gregory, Z. and Z. J. Karpowicz, H. J. Lea, A. M. MacFarlane, Lars Svensson and Alan Vittery. P. J. Hayman was the first person to advise two of us (RFP and IW) on the problems of the group and provided much help in the early stages of our study. We would also like to pay special thanks to A. R. Kitson and M. J. Helps for their dedicated support of field work carried out over the past eight years, particularly in 1966–67, and to the many others who have given us information, often as a direct result of their own field experiences and also those who supplied us with photographs, many of which form the selection published in this book.

Our thanks are also due to the staff of the Bird Room of the British Museum, where much valuable information was gleaned from the skin collection.

We are most grateful to the World Working Group on Birds of Prey, set up by the International Council for Bird Preservation, for allowing us to publish the summary of the legal status of birds of prey in Section 8.

Lastly, may we say how grateful we have been to the Editors of *British Birds* for their encouragement and guidance, particularly I. J. Ferguson-Lees and P. F. Bonham who fought with the considerable initial editorial problems when these papers were published in their journal.

Definitions

Hand	Wing between carpal joint and tip (primaries)
Arm	Wing between body and carpal joint (wing coverts and secondaries)
Soaring	Circling flight often in thermal of warm air
Gliding	Flight on a straight course without or between wing beats

Range covered by book

Although the species dealt with in this book are those in the *Field Guide to the Birds of Britain and Europe* (covering all of Europe east to 30° longitude), we have also briefly outlined known ranges in the Middle East and North Africa, for the convenience of that ever-growing band of bird-watchers who visit these regions. We have not dealt with distribution in detail, however, as it seemed desirable to put as much of the text as possible opposite the drawings and, in any case, there are many books devoted to this subject. Nevertheless, provided one exercises caution, knowing what to expect can be a great help.

Illustrations

In the main series of drawings the purpose has been to show, as accurately as possible, every important plumage phase of all the European raptor species. At the same time great efforts have been made to make the shapes accurate, and all drawings on the same page are to scale.

In addition at least one vignette drawing accompanies every species' description. It is hoped that the vignettes will convey something of the characteristic flight patterns, positions and behaviour. Lastly, a section of photographs, depicting all the species dealt with in the book, will, we believe, add authenticity to the rest of the illustrations, as well as providing much extra information. A useful tip that helps convey the feeling of seeing the birds in the field is to look at the illustrations at arm's length, or even further away, so that the points of detail become the reality of what the observer actually sees.

Whilst some of the photographs are far less sharp than we should like they are the best available for the particular identification purpose.

Introduction to Second Edition

The revisions incorporated in this second edition are the result of much further field work by ourselves (particularly Steen Christensen in Israel) and the help of a number of observers who responded to our numerous requests for 'comments and constructive criticisms' in *British Birds*. They are: Francisso Avella, D. Belshaw, R. K. Brooke, R. A. Frost, E. F. J. Garcia, P. J. Grant, G. J. Oreel, Pierre Petit, J. F. Reynolds, Peter Steyn and D. I. M. Wallace. We thank them all. Particular thanks, however, must go to Lars Svensson who has so meticulously examined our earlier texts and given us the benefit of his vast experience both in the field of identification and taxonomy. Without his tremendous help the revisions would not have been so complete.

Finally, although we have reviewed the book as a whole we have concentrated our efforts, particularly, on revising the text and illustrations of the chapters dealing with the 'buzzards' and the *Aquila* eagles; there are other significant amendments under Marsh Harrier, Eleonora's Falcon and the *Accipiters*, and over 40 new photographs. We are grateful for the continuing support of photographers who regularly send pictures of raptors in flight to us. We hope they will continue to do so.

Introduction to Third Edition

A number of refinements and additions have been made throughout the text of this edition, several new drawings have been prepared and most others redrawn or revised. The work on the buzzards and *Aquila* eagles has been particularly extensive, but all sections have been reviewed and the whole book has been reset.

In tackling these revisions we have once again been greatly encouraged by the work of Lars Svensson who has given freely of his knowledge and advice. We also wish to thank others who have made useful comments, all of which have been taken into account: A. Ash, E. J. Avellà, P. Combridge, H. Delin, Dr W. Jenning, B. King, M. Lammin-Soila, J. Parker, T. W. Parmenter, G. W. Rayner, P. Sladkovsky, W. Wagner.

In the exacting task of typing the revisions we have been much indebted to Diane Willis.

Finally, a 'thank you' to all those photographers who have sent us new material. This has enabled the plates to be more informative and has forced our publisher to increase the number from 80 to 96!

This book is intended as a practical, working guide to identification, and the authors would welcome suggestions and criticism which would be helpful when preparing subsequent editions. They should be addressed to R. F. Porter, c/o T. & A. D. Poyser Ltd.

1 The Buzzards and Honey Buzzard

In this group we are concerned with three true buzzards (Buzzard *Buteo buteo buteo*, Long-legged Buzzard *B. rufinus* and Rough-legged Buzzard *B. lagopus*), a distinctive subspecies of the first which is often confused with the second (Steppe Buzzard *Buteo buteo vulpinus*), and the only European member of another genus which is best considered at the same time (Honey Buzzard *Pernis apivorus*). These are difficult raptors because of their considerable variation in plumage and rather similar structure.

The illustrations show the under- and uppersides of each bird in various plumages and with different wing positions (while plates 1–15 emphasise some of the points in the form of photographs). Not illustrated in detail, however, are head-on profiles. This is mainly because they are too variable to be diagnostic, depending greatly on the strength of the wind and other weather conditions. As a general rule, the true buzzards soar on raised wings and the Honey Buzzard on flat wings, but variations are discussed under the individual species.

As a group the true buzzards inhabit a wide range of habitats, from tundra and steppe to forest and woodland. The Honey Buzzard is solely a woodland species, except on passage.

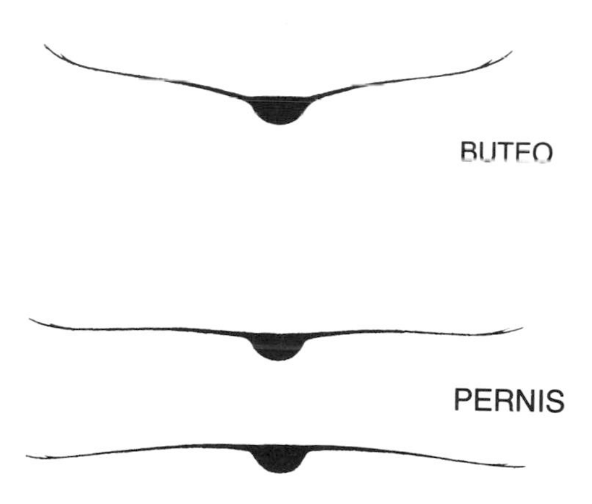

Fig. 1. Head-on soaring profiles of true buzzards *Buteo* with wings raised and Honey Buzzard *Pernis apivorus* with wings flat.

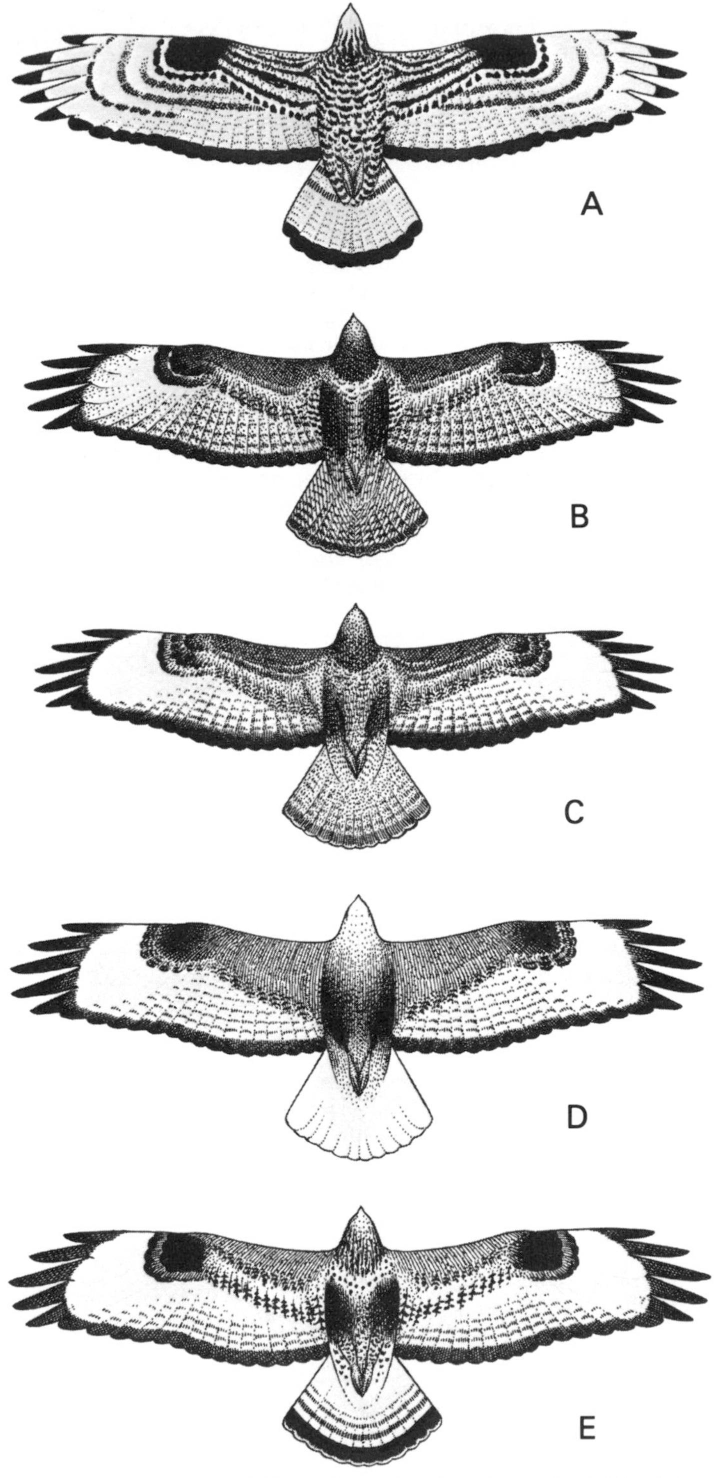

Fig. 2. Typical undersides of Honey Buzzard and the buzzards. Adults depicted.

Fig. 2A. **Honey Buzzard** *Pernis apivorus*. Adult. Slim body, small protruding head (reminiscent of Cuckoo *Cuculus canorus*), longish tail. This is normal barred type, but much confusing plumage variation. Form of bars on primaries and secondaries diagnostic and also three bands on tail; usually strongly marked carpal patches. Soars on flat wings, unlike true buzzards *Buteo* which invariably do so on raised wings. Anywhere in continental Europe, late April-September, except north Fenno-Scandia, much of Low Countries, parts of Iberia and south Italy; and on passage in Turkey, Middle East, Mediterranean area and north Africa, late April-May and August-September. Tail and wing pattern signify adult.

Fig. 2B. **Buzzard** *Buteo buteo buteo*. Adult. Compact and well-proportioned with broad head and thick-set neck; wings appear broader and tail shorter than Honey Buzzard. Plumage very variable from largely dark to largely white. Soars on raised wings. Commonest of buzzard group: anywhere in Europe, except Iceland, Faeroes, Ireland, east Britain and Scandinavia north of 62°-65°N, east to Sweden, Poland, Romania and Turkey; in Scandinavia mainly March-October because, although many populations are largely sedentary, some migrate south to Mediterranean region and north-west Africa; replaced in east Europe by another subspecies, Steppe Buzzard (see below). Clearly-defined dark band on trailing-edge of wing and subterminal band on tail indicates an adult.

Fig. 2C. **Steppe Buzzard** *Buteo buteo vulpinus*. Typical adult. Resembles Buzzard but slight structural differences give narrower-winged and longer-tailed outline. Eastern European birds often rather *B. buteo*-like in colouration though warmer and more rufous tinge, especially to tail. Wing-coverts and body vary in colouration and some show rufous underparts and orangey tail causing confusion with Long-legged Buzzard. North-east and east Europe from Fenno-Scandia and Russia south to Ukraine and Caucasus, March-October; intermediates in area from Finland to Balkans; on passage east and south Europe, Middle East and north Africa.

Fig. 2D. **Long-legged Buzzard** *Buteo rufinus*. Larger and more eagle-like than Steppe Buzzard to which it is sometimes similar in plumage. Whitish head and chestnut belly patch often characteristic but much variation. Uppertail usually whitish at base merging into light orange or rufous tip, which is normally unbarred. Juvenile has grey-brown, densely barred uppertail. Soars generally on more raised wings than Buzzard or Steppe; sluggish, often idling on telegraph poles or hummocks. In Europe only in southern Yugoslavia, Greece and Turkey, where resident, and south Russia, where migratory, but also breeds south through Middle East and across north Africa to Morocco. Adult illustrated, juveniles/immatures lacking clear-cut black band to trailing-edge of wing.

Fig. 2E. **Rough-legged Buzzard** *Buteo lagopus*. Adult. Size and structure closest to Long-legged, but head smaller. Typically shows combination of dark and white plumage with whitish underwings and white-based tail contrasting with black carpal patches, large dark area on belly and broad black sub-terminal band on tail, but some variation, particularly in underwing-coverts. Soars on raised wings; frequently hovers when hunting. North Fenno-Scandia and north Russia, April–September; winters farther south from east Britain (very scarce), Low Countries, Denmark and Germany east to Ukraine and Caucasus and south to north Italy and Balkans, October–April. Adult male depicted: note more bars on tail than female. Band on juvenile's undertail diffuse and often lacking.

Buzzard

Buteo buteo buteo
(plates 1–6, 13)

SILHOUETTE Wing-span 113–128 cm. Compact, medium-sized raptor with relatively broad wings and fairly short but ample tail. As with all true buzzards, head is short, round and comparatively broad. Fresh tail about three-quarters of width of wing and with rather sharp corners, more so than in Honey Buzzard; usually evenly rounded, but sometimes central feathers slightly elongated, making whole appear a little wedge-shaped. In head-on profile when soaring, wings normally raised; when gliding, wings flat or slightly upcurved though sometimes with hand lowered below level in the manner of Honey Buzzard. Eastern and northern-most populations of Buzzard have more slender wings and somewhat longer tail on average than central and southwest European populations, thus being closer in silhouette to Steppe and Honey Buzzard.

FLIGHT In soaring, wings raised (hand more than arm) and pressed forward; tail fanned so that sides may even reach trailing edge of wings. In gliding, arm pressed forward and hand directed clearly backwards, wings held either horizontally or with hand slightly lowered; they then look very pointed, more so than Honey Buzzard's in corresponding position. In active flight, action a little stiff and wing beats not so deep as those of Honey, possibly due to relatively shorter wings. In some parts of range, but not in all, frequently hovers when hunting.

IDENTIFICATION May be confused with many medium-sized raptors, owing to great variety of plumages, intermediate size and not very characteristic flight silhouette (though soaring on raised wings is good guide). Easily confused with Honey Buzzard, but is noticeably broader-headed and shorter-tailed and appears broader-winged, while flight actions and head-on profiles are different. Buzzard also lacks Honey's three tail bands (though these are visible only at close range or when tail spread) and distinct bars on underwing. Dark Buzzards may be confused with dark-phase Booted Eagles *Hieraaetus pennatus*, but never show the latter's combination of all blackish-brown secondaries and lighter inner primaries on underwing. Light U on breast and pale bar on median underwing-coverts of Buzzard not seen in dark Booted Eagle. Dark individuals may also be confused at distance with young Marsh Harriers *Circus aeruginosus* but Buzzard has shorter tail and broader wings. Some pale Buzzards much resemble Rough-legged Buzzards, showing whitish base to uppertail, dark belly patch and large blackish carpal patches, but see that species for distinction. Other pale Buzzards look like Short-toed Eagles *Circaetus gallicus*, owing to combination of dark head, blackish wing-tips and otherwise very light undersides, but Short-toed is much larger with comparatively much longer wings, broader and more protruding head and quite square-cut end to tail which has very sharp corners. Some whitish Buzzards may resemble pale phase Booted Eagles (which see), but underside of Buzzard's flight-feathers are always more or less dark-barred on pale ground colour, whereas in Booted they appear all-blackish with pale inner primaries, if seen well. Can also be confused with Steppe Buzzard, and some eastern Buzzards are impossible to separate from western Steppe Buzzards, which is not surprising since they are conspecific and interbred in eastern Europe. In general, Buzzard is most safely identified by size, compact flight silhouette, comparatively short, dark and evenly barred tail, and broad wings raised when soaring.

Fig. 3. **Buzzards** *Buteo buteo buteo* from above. Upperparts vary as much as underparts and some strikingly marked creamy-white and blackish-brown birds occur though these are fairly unusual (eg F, G, H). The most typical plumages are A, B and D with lighter-patterned birds occurring more often in northern parts of range. Some typical nominate birds show light area on primaries (eg B) as seen in Steppe Buzzards, this frequently being a feature of British birds and intermediate forms from eastern and northern parts of range. (Palest birds after Lars Svensson).

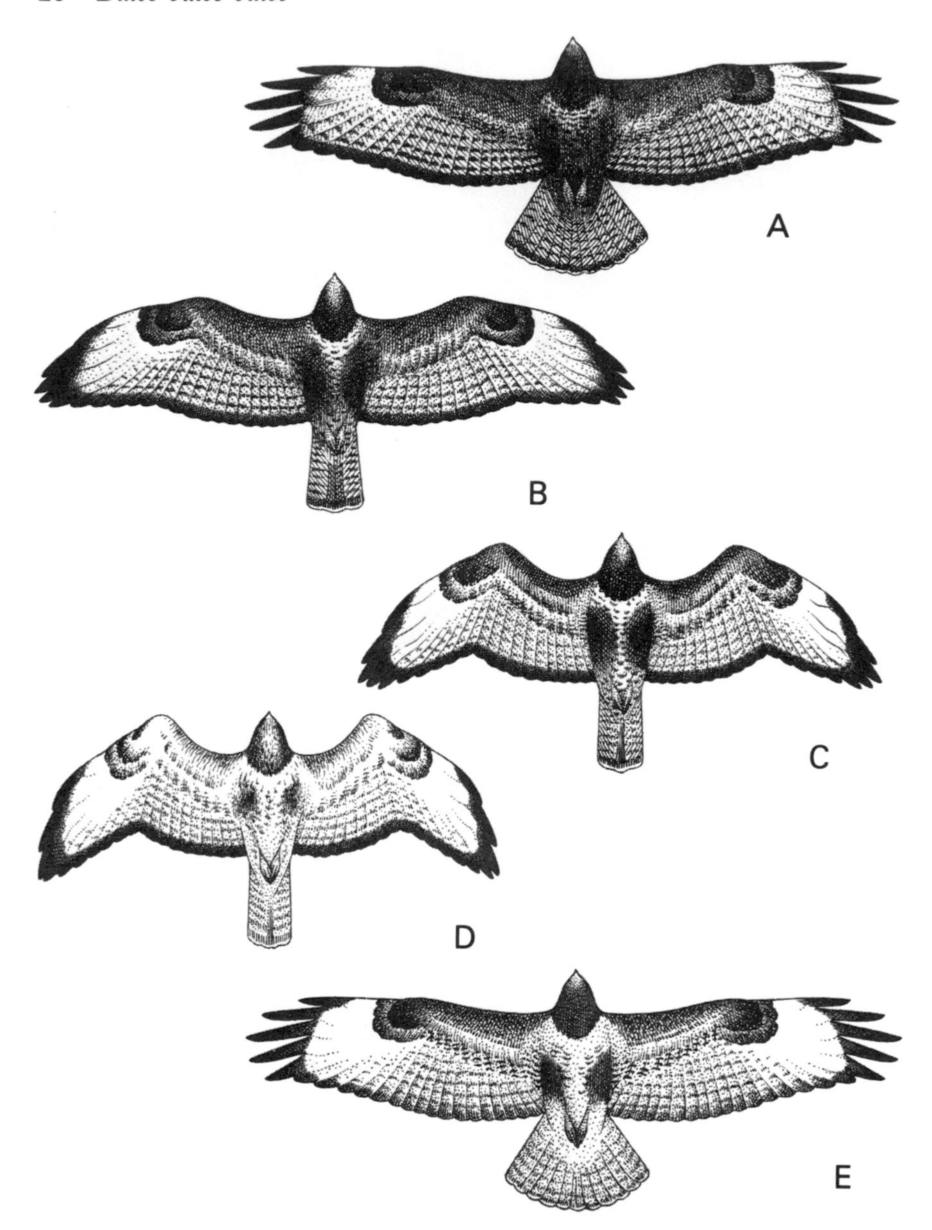

Fig. 4. **Buzzard** *Buteo buteo buteo* from below. Most variable of all Palearctic birds of prey; illustrations show range of plumages that can be encountered. Variations are greatest in north European populations which can be any shade beneath from blackish-brown with paler flight feathers (A) to almost creamy-white (K) with complete range of barred, streaked and blotched individuals in between. (Lightest birds are most likely to be confused with Short-toed Eagle.) In western and central Europe darker types (A and B) are commonest. Note presence of light U on breast and pale bar on median underwing-coverts which is nearly always present even on darker birds (A–F). No apparent correlation between plumage and sex but adults (A–D) may be told by clear-cut blackish-brown band along trailing-edge of underwing and a broader subterminal band to tail. In juveniles (E–I) dark subterminal tail bar is same width as, or hardly wider than, other bands, whilst dark band along rear edge of wing is noticeably paler than in adult and much more diffuse along inner edge. (Palest birds after Lars Svensson).

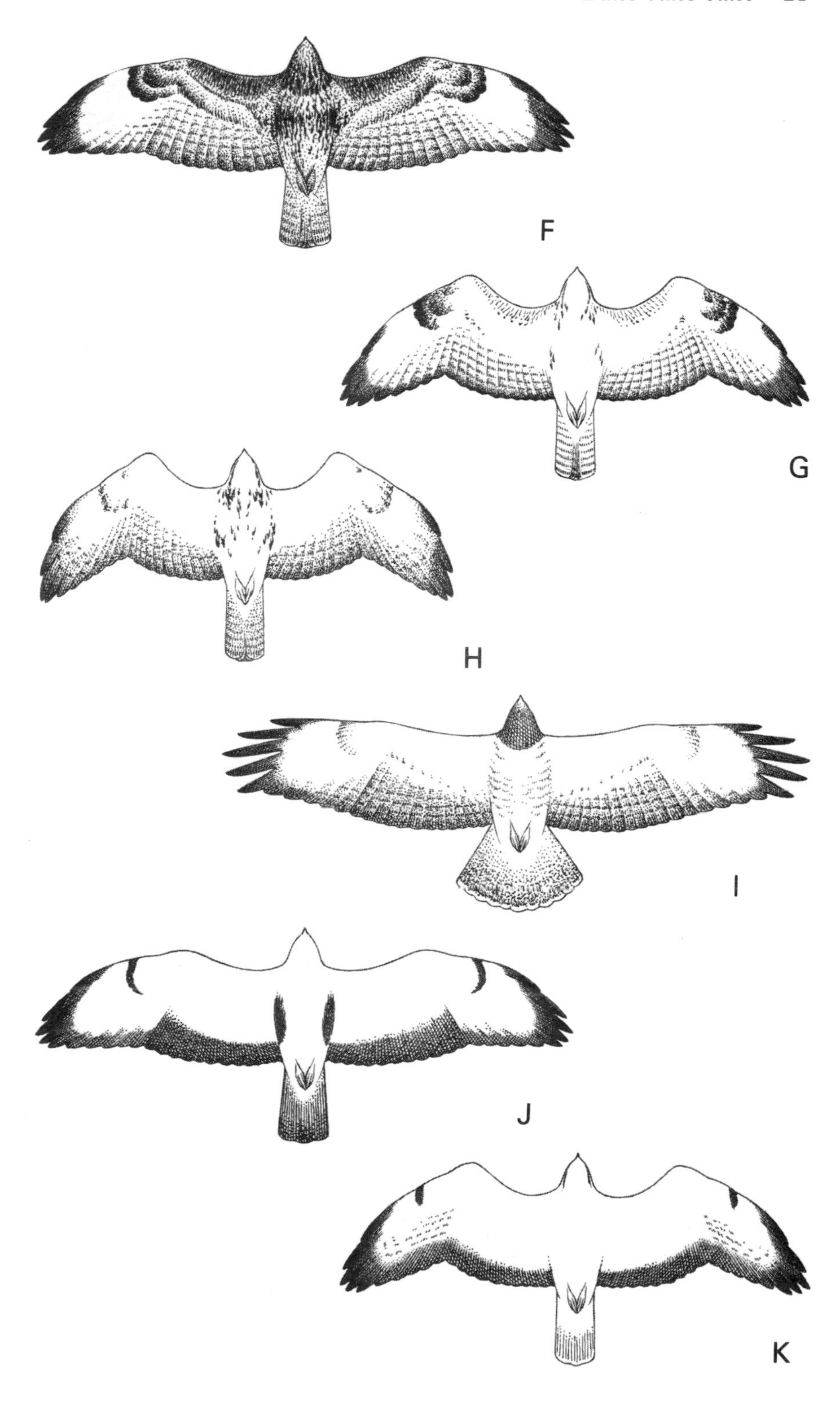
F
G
H
I
J
K

Steppe Buzzard
Buteo buteo vulpinus
(plates 2, 4, 6)

SILHOUETTE Wing-span 113–128 cm. This eastern subspecies averages a fraction smaller than the Buzzard, but individual and sexual variation make size quite useless for identification in the field. Wings slightly slenderer than those of Buzzard and, as a result, tail appears a little longer, the whole effect being to produce a shape not unlike some Honey Buzzards. Otherwise like Buzzard, including head-on profile.

FLIGHT Identical to that of Buzzard, though active flight slightly less clumsy and wing-beats a little faster.

IDENTIFICATION In Europe three phases of Steppe Buzzard occur: a dark phase, a medium phase and a fox-red phase. Dark and medium phases resemble Buzzard with ground colour varying from brown, rusty brown to rufous-ochre. Barred tail either greyish, brownish or rufous. Adults of both phases have similar underparts to Buzzard and it can be very difficult to distinguish between the two subspecies, especially amongst intermediate populations in region stretching from Finland to the Balkans. Fox-red phase is easier to distinguish from Buzzard by its more or less uniformly fox-red-coloured body and wing-coverts; easily confused, however, with Long-legged Buzzard (which see), though smaller with comparatively shorter wings (and clearly smaller wing span) and has less deep and elastic wing-beats, and different head-on wing position when gliding. This phase rare west of 30°E, western limit of breeding range, though many can be observed at Bosphorous on autumn passage. Long-legged also usually has whiter head and breast and much darker belly, as well as more contrast on upperside between rump and tail and between wing-coverts and flight-feathers; in addition, pale patches on upper hand average larger and paler; tail often lacks any trace of barring, which is rarely the case with Steppe Buzzard, at least when seen at close range. Pale Steppe with dark on breast may superficially resemble Rough-legged, but comparatively shorter-winged, has different head-on profile, less elastic wing-beats and lacks broad, conspicuous dark band which that species always shows on upper- (and, in adults, under-) tail. Uniform blackish-brown phase has not been recorded in Europe. Its plumage is identical to the dark phase of Long-legged Buzzard, which similarly has not been recorded in Europe. Both have been recorded on spring migration in the Near East.

(*opposite*)

Fig. 5. **Steppe Buzzard** *Buteo buteo vulpinus*. From above. A and B adults of fox-red phase; C adult and D juvenile of brownish, Buzzard-like phase; and E adult of blackish-brown phase. The majority of adult Steppe Buzzards have rather uniform rusty-brown (brownish phase) or rufous-reddish (fox-red phase) upperparts, slightly darker secondaries with dark band to trailing edge and often conspicuous pale primary patch. Brownish phase adults have grey, brownish or rufous tail densely barred with wide dark subterminal band; adults of fox-red phase have orangey-red to brownish-red tail, sometimes paler at base and with variable barring (see A and B). Juveniles of fox-red phase have usually completely barred tail without the wide subterminal band. Some fox-red Steppe Buzzards in Near East have pale upperwing-coverts and head and this, together with orangey-, almost white-, based tail, makes for confusion with Long-legged Buzzard. The blackish-brown phase (E) has uniform blackish-brown body and wing-coverts and, in adults, wide blackish band on trailing edge of wing and near tip of completely barred tail; its plumage is identical in the field to similar phase of Long-legged Buzzard and is distinguished by silhouette and flight.

A
B
C
D
E

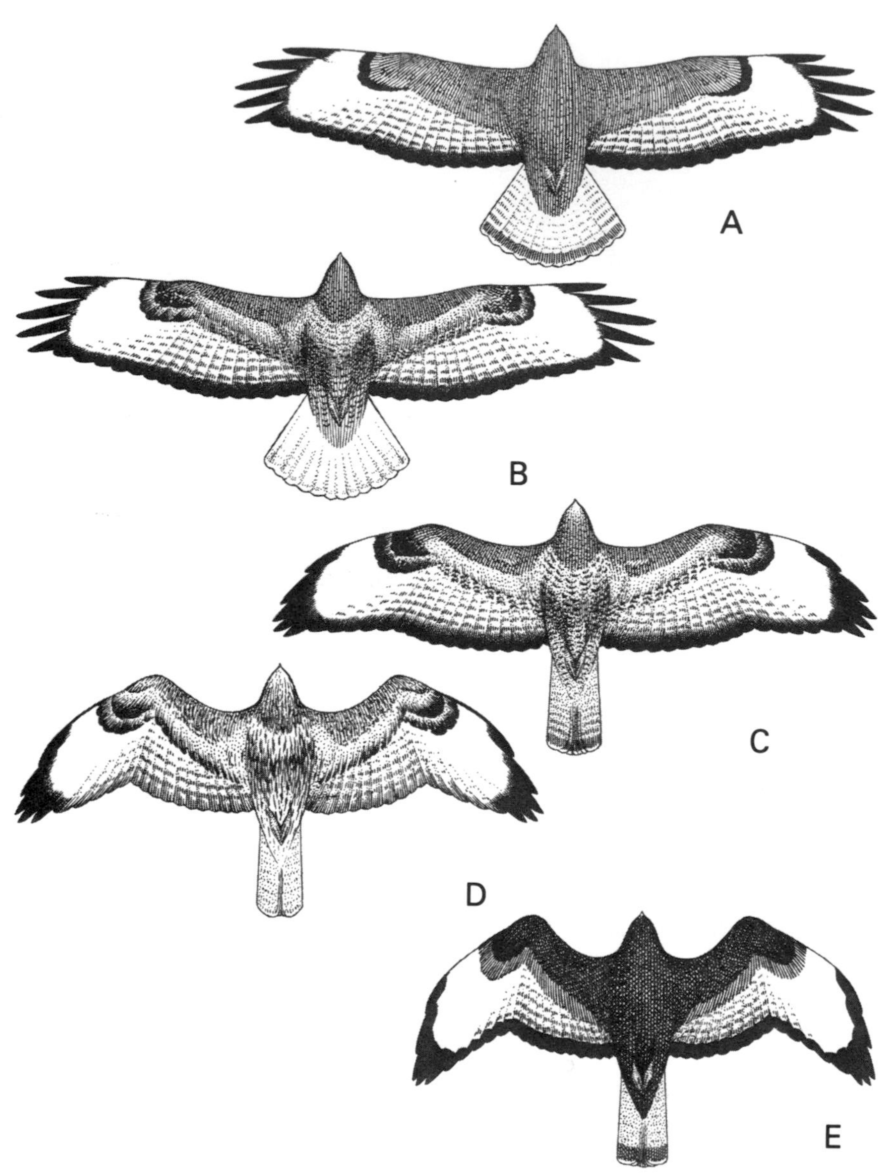

Fig. 6. **Steppe Buzzard** *Buteo buteo vulpinus*. From below. A and B adults of fox-red phase; C adult of brownish phase; D juvenile; E adult of blackish-brown phase. Less variable than nominate race. In fox-red phase (A and B) body and underwing-coverts are fox-red, either uniform (A) but often with reddish barring below breast and paler bar on median coverts (B). In brownish phase (C) pattern very like Buzzard, but ground colour often more rufous-brown. Flashing white flight-feathers, particularly primaries, bordered by usually clear-cut broad blackish band along trailing edge, characteristic of most Steppe Buzzards. Tail generally paler than Buzzard; in fox-red phase whitish to creamy, or even pinkish, either barred throughout with broad blackish subterminal band (A) or virtually unbarred (B). Adult of blackish-brown phase (E) has rather broad and clear-cut blackish band along trailing edge of wings and rather broad blackish subterminal tail band and rest of pale grey tail finely barred throughout; body and wing-coverts uniform blackish-brown, like corresponding phase of Long-legged Buzzard. Juvenile (C), which lacks clear-cut blackish-band to trailing edge to wing, very similar in plumage to juvenile Buzzard and virtually impossible to separate in field except in fox-red phase.

Long-legged Buzzard
Buteo rufinus
(plates 7–9)

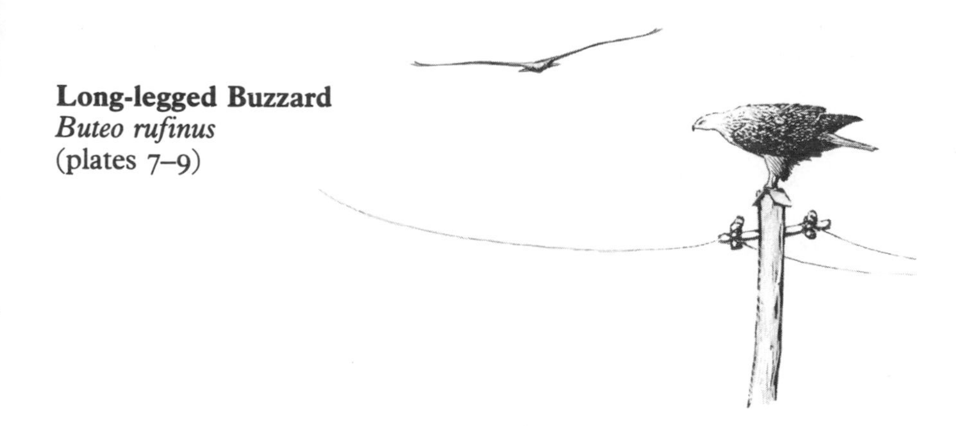

SILHOUETTE Wing-span 126–148 cm. Size similar to Rough-legged Buzzard (compare Figs. 2D and 2E), but wing-span sometimes slightly larger. Head typical of true buzzards, being short, round and broad. Wider-winged and relatively still longer-winged and especially longer-armed than Buzzard and Steppe Buzzard; wings also often look almost rectangular, hand being only slightly narrower than arm. Ample tail comparatively long in relation to Buzzard and Steppe Buzzard, corresponding to greater width of wing; usually slightly rounded or tapering, but sometimes looking square-cut. In head-on profile when gliding, arms are slightly lifted and hands level producing a bend like in Rough-legged Buzzard. When gliding slowly (sailing) wings are normally flat and almost straight out from body with a straight rear edge, a slightly angled leading edge and only the tips pointing backwards. When soaring wings lifted well above level, generally more than Buzzard.

FLIGHT In soaring, wings are somewhat lifted and pressed slightly forward. When gliding slowly, in striking contrast to the soaring position, wings are normally carried flat and almost straight out from the body with a straight rear edge, a slightly angled leading edge and only the tips pointing backwards. Active flight involves elastic and consistent wing-beats. When hunting it glides, sails and soars, often interrupted by hovering like Rough-legged Buzzard.

IDENTIFICATION Not difficult to distinguish from Buzzard and Honey Buzzard, but easily confused with Steppe Buzzard and Rough-legged Buzzard. Long-legged must often be separated from Steppe by its outline, being somewhat larger and especially longer-winged with slower, more elastic and consistent wing beats and a different head-on profile when gliding (see above). If seen with Steppe Buzzard, identification on size is fairly easy, but plumages rather similar. Even so, contrast between dark belly and remaining pale underside is usually a good field character, though few Long-legged Buzzards almost or entirely lack the dark patch on belly. Though exceptions occur, head is much paler than that of Steppe, underparts show greater contrast between coverts and flight-feathers, pale patch on upper hand averages larger and paler, thus also making primaries more translucent and finally, blackish carpal patches below average considerably larger. The tail of Long-legged Buzzard averages paler (especially at base) and is less barred, but some typical Steppes have tail very like Long-legged, making the character dubious. Size, flight silhouette and flight actions also very similar to those of Rough-legged Buzzard, though Long-legged sometimes looks slightly longer-winged. Pale unbarred tail of adult Long-legged is best identification mark in this case and even immatures with barred tails do not have the wide terminal band of Rough-legged. Rest of upperparts rather similar in these three species, though there tends to be a somewhat greater contrast between flight-feathers and wing-coverts in Long-legged, especially in worn plumage. Patterns of underparts vary and, though typical reddish-brown individuals can easily be separated from Rough-legged, some Long-legged have such pale underwing-

(*continued p. 28*)

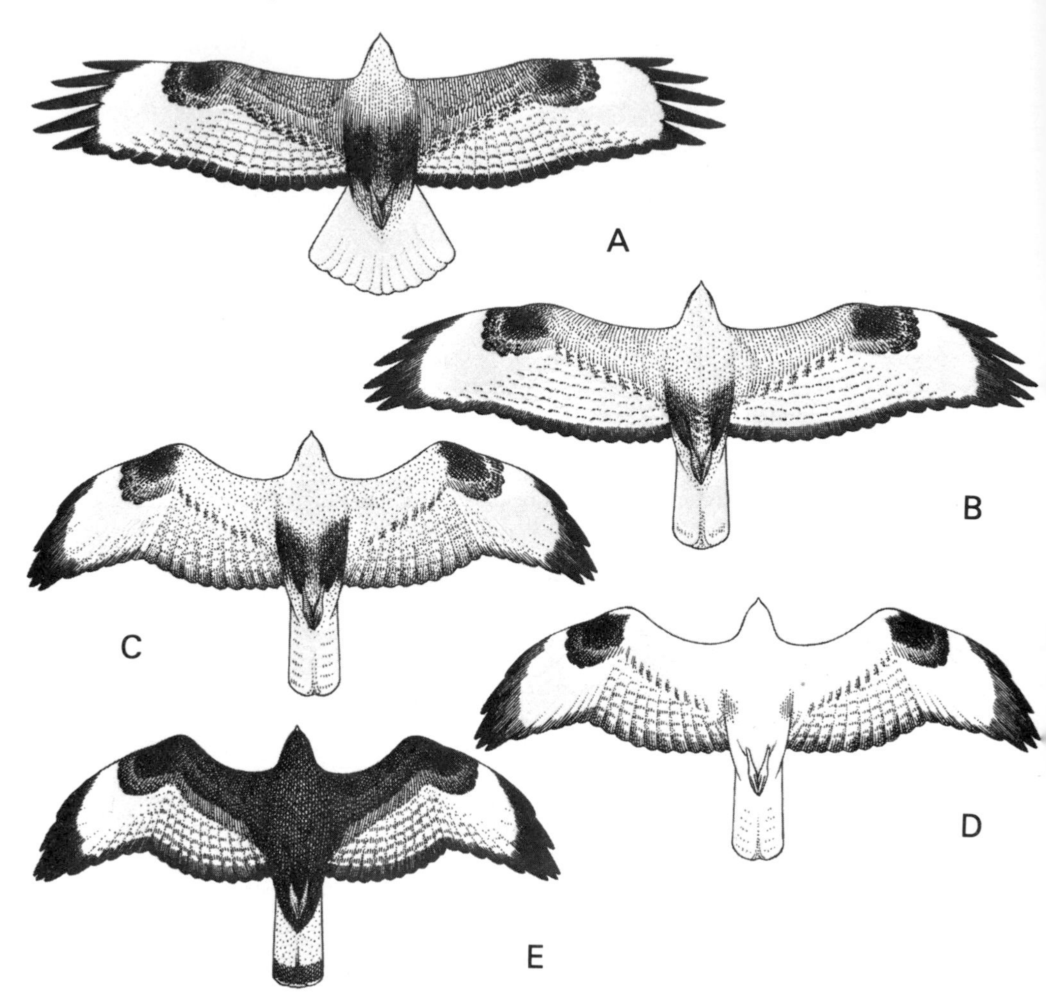

Fig. 7. **Long-legged Buzzard** *Buteo rufinus*. From below. A and B adults of medium and pale phase (commonest type); C typical juvenile and D juvenile of rare whitish phase; E adult of blackish-brown phase. Varies almost as much as Buzzard, but certain characters are more constant. Typical Long-leggeds have almost creamy-white head and lower-neck with pale reddish-brown breast and dark brownish belly and 'trousers' (A); or with creamy-white extending onto breast and causing more clear-cut demarcation from brown belly (B). However, a few Long-legged Buzzards lack any conspicuous dark colour on belly area or just show dark on sides of belly/'trousers'. Underwing-coverts either reddish-brown (A) or pale creamy-buff (B) or intermediate. Some individuals are very pale, almost snow-white below, and (D) depicts juvenile example. In all phases primary-coverts are usually blackish, forming large carpal patch. Flight-feathers, particularly the translucent primaries, are strikingly white in adults (A, B and E) with broad blackish clear-cut band along trailing edge of wing. When spread, translucent tail is creamy-white below, though often looks orangey, due to light shining through translucent feathers which are this colour above. Tail normally unbarred in adults (A), though some show faint traces near tip (B). In typical juvenile (C) flight-feathers lack clear-cut blackish band on trailing-edge of wings; band being blurred and diffuse as in young Buzzards; pale undertail faintly barred throughout (especially noticeable in spread tail) and body feathers lack tendency of cross-bars, seen in some adults. Juveniles, otherwise, patterned as adults. Adult blackish-brown phase (E) is uniform dark chocolate brown or blackish-brown on body and wing-coverts, has broad blackish band along trailing-edge of wings and greyish-white undertail has very broad blackish subterminal band; when spread, tail shows faint dark barring.

Fig. 8. **Long-legged Buzzard** *Buteo rufinus*. From above. A and B adults; C juvenile and D adult of blackish-brown phase. Head and neck usually creamy-white or pale rufous; mantle and wing-coverts pale rufous-brown (A) or yellowish-brown (B), contrasting with grey-brown flight-feathers. Dark brown rump contrasts with usually rather pale upper-tail, which in adult bird is an unbarred light orange, merging into an almost whitish base (B), or more uniform pale red-brown tail, only slightly paler at base and sometimes with faint subterminal bar (A). Juvenile (C) has darker grey-brown tail barred throughout; immatures have reduced tail-barring. A large greyish-white primary patch visible at all ages though variable in size and distinctness. Adult blackish-brown phase (D) is uniform blackish-brown or dark chocolate-brown all over, apart from pale primary patch and tail which is dark-barred throughout with a broad blackish subterminal band.

coverts that the only difference that is at all constant, apart from the tail, is the paler breast of this species: Rough-legged often has rather dark streaks on the breast, these separated from the still darker belly patch by a whitish U. Even this is not always clear, however, and sometimes the only safe distinction from Rough-legged is the unbanded tail. The melanistic phase may be recognised from other Buteos by silhouette and striking white primaries below. However, a few of the dark phase Steppe Buzzards are so similar to dark Long-legged Buzzards that we know of no way to separate them if the two species are not seen together and comparison of size and proportions made. This problem should not arise in Europe but will do so in the Middle East.

Rough-legged Buzzard

Buteo lagopus
(plates 10–12)

SILHOUETTE Wing-span 120–150 cm. Slightly larger and relatively longer-winged than Buzzard (compare Figs. 2B and 2E): difference in wing span often considerable, though not always evident. Head typical of true buzzards, being short, round and broad. Wings relatively narrow, rear edge slightly curved and hand little narrower than arm. Ample tail comparatively long, corresponding to width of wing or little longer; tip slightly tapering or rounded.

FLIGHT Slow, easy and consistent with loose, not very deep wing beats, varying in number and interspersed with glides on very slightly raised wings. The general appearance being not unlike Hen Harrier *Circus cyaneus* or Honey Buzzard. Frequently hovers with wings held almost motionless (arm raised, hand flat and primaries slightly upturned), head down and peering about and tail constantly manoeuvred in a Kite-like fashion. Soars with wings raised and pressed forward. In gliding, arms raised while hand level.

IDENTIFICATION Often confused with Buzzard and Long-legged Buzzard. Indeed, owing to long wings, large individuals may be mistaken for medium-sized eagles, though distinguished by shorter neck, smaller bill and narrower wings raised in soaring; wing-tip also has only four, not six, free primaries ('fingers'). Pale Buzzards and also Honey Buzzards may show white uppertail-coverts and basal tail-feathers, all-white underwing with black carpal patch and even dark area on belly, but tail is seldom so sharply black-and-white as in Rough-legged, white on base of rectrices normally being limited to area adjacent to upper-tail-coverts and often to central tail-feathers only; in addition, Rough-legged seldom shows creamy-white patches on upperwing-coverts and shoulders, often seen in pale Buzzard and Honey. Few Buzzards have pale patch on primaries above, so often helpful when present in Rough-legged. Differences in silhouette, particularly in active flight, are useful. Elastic wing beats of Rough-legged quite unlike those of stiffer-winged Buzzard. Head-on soaring profile often diagnostic: wings are slightly raised in Buzzard, arm raised and hand level in Rough-legged. Confusion possible with pale Steppe, which may show whitish body and underwing, with black carpal patch and trailing edge, often also dark area on breast and often pale patch on primaries above; normally, Rough-legged can be identified by its broad black tail-band, but some Steppe have pinkish or whitish tail with dark terminal band as only visible barring. So advisable to look for other characters, notably size, head-on profile and active flight method; Rough-legged also has white leading edge to arm, sometimes conspicuous. Apart from black on tail, Rough-legged difficult to tell from some Long-legged, mainly pale ones, particularly as differences in flight and outline are slight and inconstant. But adult Rough-legged with dark throat and whitish U on breast quite distinct from Long-legged. In general, wide black tail-band is by far the best identification feature of adults, while the dark greyish band on otherwise unmarked pale tail is diagnostic of young birds.

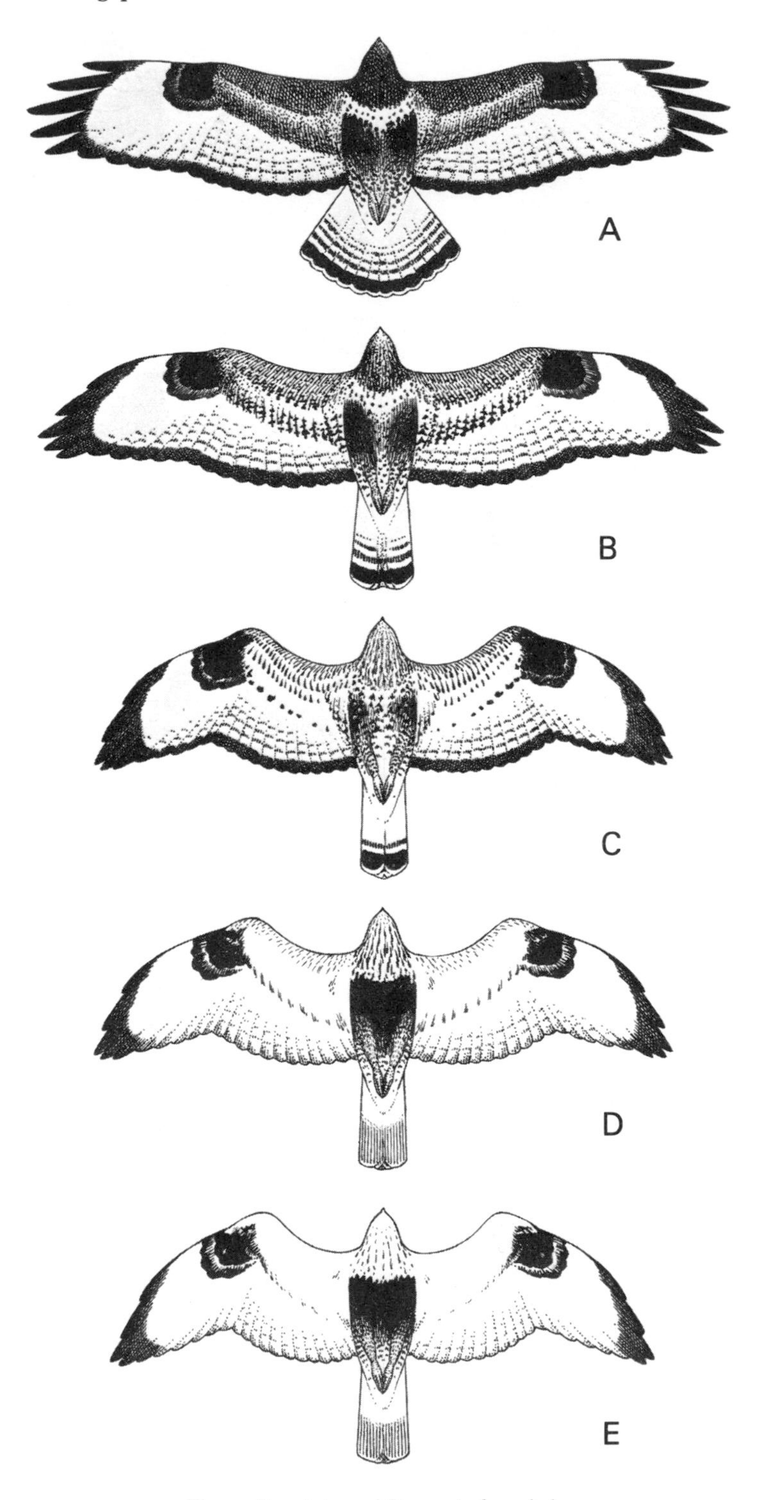

Fig. 9. Rough-legged Buzzards from below

Fig. 10. **Rough-legged Buzzard** *Buteo lagopus*. From above. A and B adults, C juvenile. Head and nape are whitish with variable dark streaking. Juveniles average paler on mantle and upperwing-coverts (particularly pale in this individual), which are patterned with greyish-brown to a greater or lesser extent, with the median-coverts generally the palest area. Sometimes there are whitish feather edgings to the coverts, particularly greater-coverts. The juvenile sub-terminal tail bar is wider, lighter and less clear-cut than that of adult. Upperparts of adult average darker and show less contrast to flight feathers. Tail has 2–4 clear-cut black bars, on greyish or white background and white on tip of tail wider than in juvenile. Most Rough-legged Buzzards show distinct greyish or whitish patch to primaries, in juveniles these can be so well marked as to recall young Golden Eagle *Aquila chrysaetos*. A whitish leading edge to arm is constant in all ages. Because of individual variation age is best determined on tail markings. Adult males have the most bars.

(*opposite*)

Fig. 9. **Rough-legged Buzzard** *Buteo lagopus*. From below. Underparts rather variable and adults (A–C) average considerably darker than juveniles (D–E) with darker blackish-brown throat, blackish or dark greyish-brown densely spotted underwing-coverts, often mixed with rusty-brown, heavier blacker barred flight-feathers (mainly secondaries) with clear-cut blackish band along trailing-edge of wings. There are 2–4 narrower, more well-marked black tail-bars, the subterminal widest. As a result of normally dark throat in adults, a more or less distinct whitish U on breast is more prominent in adults and blackish-brown patch on lower breast/belly often shows up as a patch on each side of upper belly. Juveniles (D and E) average clearly paler, sometimes snow-white on entire underparts, except for dark belly-patch (more often covering whole of belly), large blackish carpal patches, black wing-tips and without broad, clear-cut band along trailing edge of wings. Single subterminal tail-bar of juvenile bird is wider, and much paler than adult, often whole undertail appearing whitish. Much individual variation in the dark markings on throat, underwing-coverts and flight-feathers, and age determination sometimes very difficult if tail-markings are not seen. Translucent primaries most evident in birds with prominent pale patch on primaries above. Odd birds are completely without dark belly-patch but this is most exceptional. Immatures (2nd year with second generation of tail- and flight-feathers) are generally paler below than adults and show gradation between juvenile and adult plumage.

Honey Buzzard
Pernis apivorus
(plates 13–15)

Honey Buzzard in display flight

SILHOUETTE Wing-span 135–150 cm. Size similar to Buzzard, though wing-span sometimes a little larger (compare Figs. 2A and 2B). Slender cuckoo-like head protrudes markedly and appears almost to form right-angle with wings. Wings proportionately narrower than Buzzards (though difference often insignificant) with little longer and more ample hand, comparatively shorter arm, and rear edge right-angled and often also characteristically pinched in at body. Most adults have smoothly curved trailing edges to wings when soaring, often appearing almost parallel edged in wings, though exceptions occur. Despite some variation, juveniles of the year have most secondaries slightly longer than outer secondaries and primaries, less frequently seen in adults. Tail comparatively long, corresponding to width of wings or even longer, though young juveniles have somewhat shorter tails; sides of tail slightly convex, tail-corners rounded and central rectrices usually slightly shorter than next innermost, producing a slight but often characteristic notch at end of tail. The spread tail looks less rounded than in Buzzard and at close quarters the shorter outermost rectrices are detectable. However, due to wear, the shape of end of tail is sometimes a dubious character.

FLIGHT In soaring, wings almost straight out from body and flat, sometimes with hand slightly lowered or even lifted and pressed clearly forward; edges usually nearly parallel (but see above). In slow gliding, wings angled with carpal joint clearly pressed forward and long ample hand pointing backwards and the rear edge characteristically right-angled to body. In active flight, wing-beats are soft, deep and elastic with undulating wave-motion of wings. Does not hover.

IDENTIFICATION May be confused with Booted Eagle *Hieraaetus pennatus*, Rough-legged Buzzard and especially Buzzard, though first two easily distinguishable on plumage, let alone silhouette. For example, Booted has blackish flight-feathers apart from pale inner primaries. Distinctions from Rough-legged given under that species: in particular, the differently shaped tail of Honey is never sharply black and white and subsidiary bands at base are reasonably conspicuous. Turning to Buzzard, it is often possible, despite great variations in both species, to distinguish adult Honey on plumage, it usually being more strongly and regularly barred underneath with barring on flight-feathers characteristic; it has a generally broader black clear-cut band along trailing edge of wings (only this broad in minority of Buzzards) and always 1–2 narrower (but still fairly broad) blackish bands across bases of outer secondaries and primaries (lacking in Buzzard), and black on wing-tip confined to just the outer tips. When visible, pattern and space between tail bands in Honey is diagnostic. Honey also lacks light U on breast seen on majority of Buzzards. Juvenile Honey is much more difficult to distinguish on plumage, and dark eye and yellow cere (yellow eye and dark cere in adult) recall Buzzard. Many young Honey Buzzards have dark secondaries below, contrasting more with pale primaries than in Buzzards; barring on flight-feathers is generally more pronounced, extending onto primaries, and some dark young Honey Buzzards show diffuse pale band on underwing formed by pale greater coverts, separating darkish secondaries from remainder of dark wing-coverts. When visible, the four dark tail-bands of juveniles are characteristic. Silhouette in flight is very useful: Honey has more protruding and narrower head and neck, comparatively narrower wings and longer, differently-shaped tail (see SILHOUETTE). Active flight involves softer, more elastic, deeper and wave-like wing-beats, while Buzzard is stiffer-winged. In gliding, Honey keeps arm more right-angled to body and longer, more ample hand is pressed more directly backwards. Buzzard glides on flat or curved wings and soars with them raised; during strong winds, however, even these differences are less reliable. Shape of Honey's longer tail is normally good guide provided it is not too worn. Though few Honey Buzzards may be difficult to distinguish from Buzzard (particularly juvenile birds with shorter tail and less characteristic pattern to flight-feathers below), observers should be able to separate them on mode of active flight at distance with experience.

In breeding season, display flight of the Honey Buzzard is a good character. After a long glide it rises steeply and, slightly hovering, 'shakes' wings above back in rapid succession, usually six or seven times. This performance repeated several times during display flight.

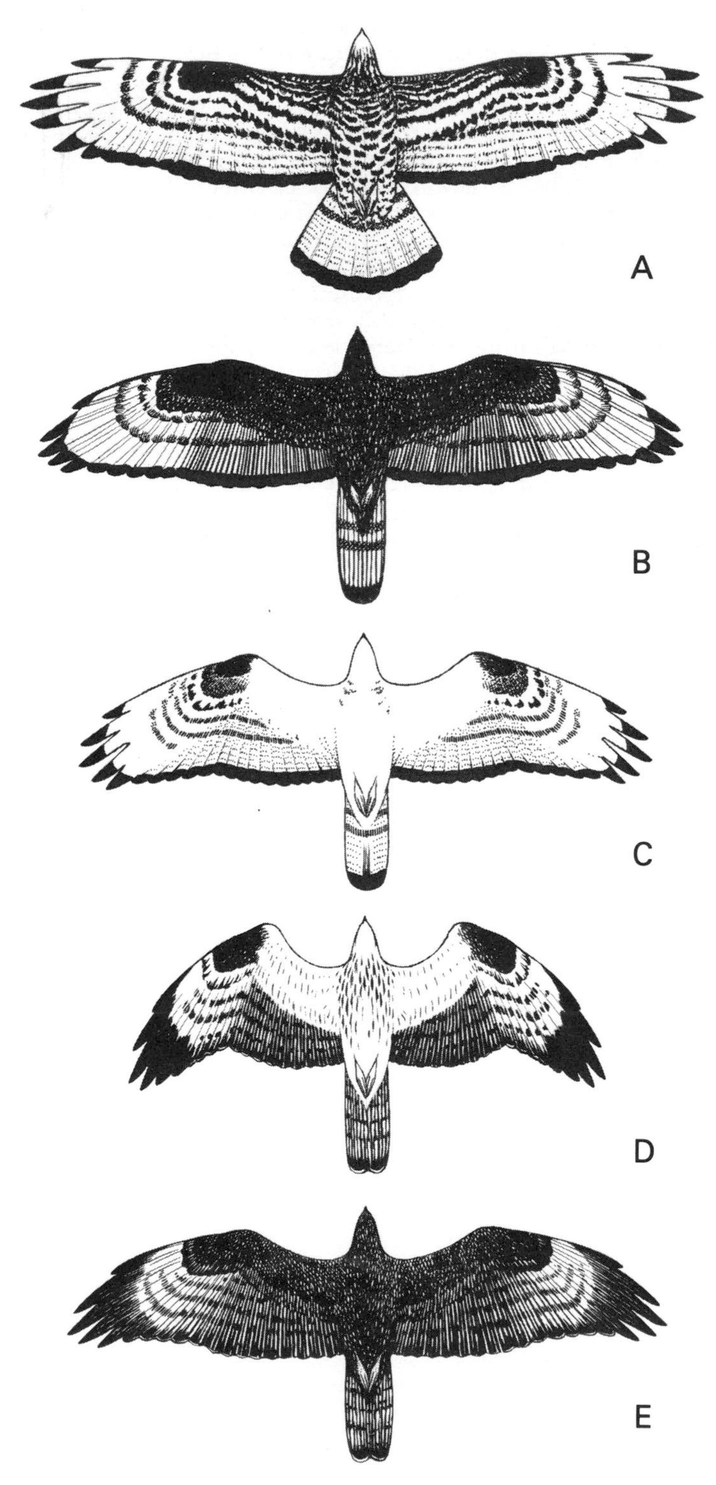

Fig. 11. Honey Buzzards from below

Fig. 12. **Honey Buzzard** *Pernis apivorus*. From above. Adult (A); juvenile dark type (B); juvenile pale type (C). Adult bird is normally greyish-brown on wing-coverts and back, but dark phase adults are usually darker above, with females more brownish than greyish. Upperhead is bluish-grey in males, less so or greyish-brown in females, but variations occur. Tail is greyish-brown with three characteristic black bands. Greyish-brown flight-feathers are bordered by blackish trailing edge; dark tips to greater-coverts, and translucent primaries show dark band at base.

Commonest juvenile type (B) has dark brown head, back and most wing-coverts; secondaries are all dark and the pale primary-patch is raggedly barred. Thin whitish line on trailing-edge of wings and tail, on tips of greater-coverts and whitish upper tail-coverts all typical of juvenile as are usually four tail bands (less noticeable than adults three). Pale juvenile (C) has white head, black eye-mask and more or less whitish, dark mottled forewing-coverts.

Fig. 11. (*opposite*). **Honey Buzzards** from below. Adults A, B, C and juveniles D, E. The typical barred adult (11A) has whitish body and underwing-coverts variably barred, and greyish-white flight feathers very faintly barred with greyish and with distinct broad, blackish bar to hind wing, and usually distinct bars to base of flight feathers often confined to outer secondaries and primaries. Carpal patches usually dark. Underside of tail greyish-white with characteristic three bars. Dark individuals (11B) have more or less uniform dark brown body and underwing-coverts, flight feathers vary, being either dark greyish on secondaries and pale greyish-white on primaries or all pale greyish-white, both with typical barring. Intermediates between 11A and B frequently occur, as do very pale cream-coloured or white individuals (11C). These may superficially resemble Rough-legged Buzzards or Short-toed Eagles. The translucent primaries are characteristic in all adult plumages. Juveniles (11D, E) differ distinctly in having yellow cere (blue-grey in adult) and dark eye (not orange-yellow iris of adult). Whether juvenile is whitish (D) rusty-brown or blackish-brown (E)—the commonest type—it differs in having darker tail with about 4 less noticeable tail bars; it lacks well-marked blackish bar to hind wing and base of flight feathers, instead all or most of secondaries are dark grey or blackish-grey more or less clear-cut from paler greyish-white primaries; all secondaries and inner primaries show 3–4 ragged bars of similar width. Blackish-grey on wing-tips more extensive in juveniles. In many dark juveniles palish bar on greater-coverts below, separating dark secondaries from dark underwing-coverts, seems characteristic. White-headed birds are found among juveniles, many having a characteristic black eye-patch. Juveniles in spring of second year have more cross-barred, adult-like body and underwing-coverts but still retain juvenile flight-feathers which by this time are much bleached, even dark wing-tip.

2 The Eagles and Osprey

Some of the species in this group are equally as difficult to identify as those of the buzzard complex, in particular the two spotted eagles *Aquila clanga* and *A. pomarina*, and Imperial and Steppe Eagles *A. heliaca* and *A. nipalensis*. All are large, except for the Booted Eagle *Hieraaetus pennatus* with its rather kite-like proportions. Amongst this group the *Aquila* eagles provide the observer with the most headaches. They may be divided into three groups according to habitat. Firstly, the mountain group, represented solely in Europe by the Golden Eagle *A. chrysaetos*, a very large species characterised by raised wings when soaring and to a lesser extent when gliding. Second, the open country group, with Imperial and Steppe Eagles, two large species which usually soar and glide on flat wings. Third, the woodland group with Spotted and Lesser Spotted Eagles, two smaller species which hold their wings down-curved in gliding and less obviously so when soaring.

All the *Aquila* eagles show great variation in plumage but, unlike the 'buzzards' this is related primarily to age which can often be determined approximately in the field. Unlike most other birds of prey, *Aquila* eagles are more contrasted and more easily

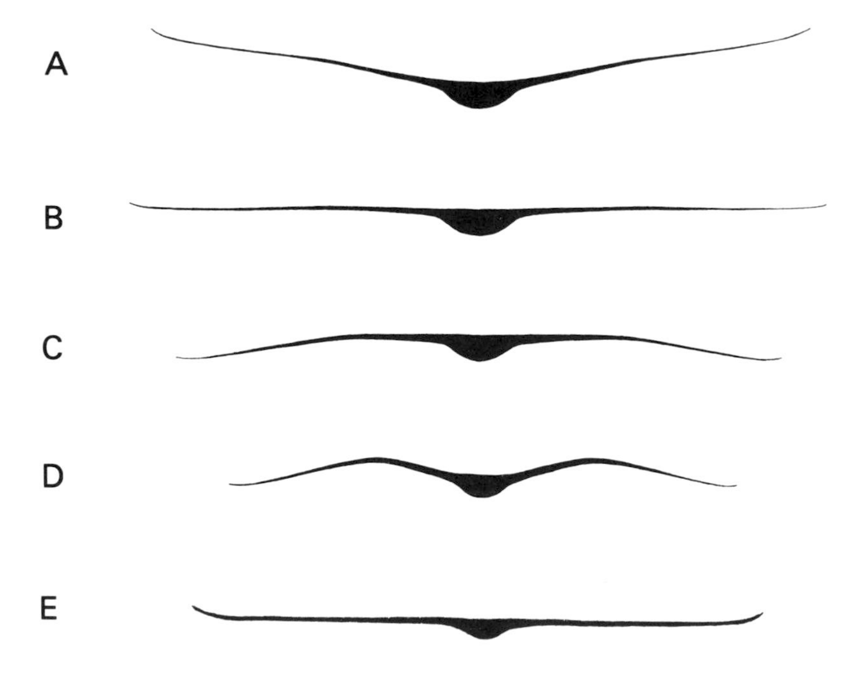

Fig. 13. Head-on profiles of eagles: (A) soaring Golden *Aquila chrysaetos*; (B) soaring/gliding Imperial *A. heliaca*, Steppe *A. nipalensis* and White-tailed *Haliaeetus albicilla*; (C) soaring/gliding Spotted *A. clanga* and Lesser Spotted *A. pomarina*; (D) Osprey *Pandion haliaetus* and (E) Booted and Bonelli's Eagles *Hieraaetus* and Short-toed Eagle *Circaetus gallicus* (but see text for details).

identified in juvenile and other immature plumages, all becoming more or less brown as adults. On the other hand, the only visible difference between the sexes is the slightly larger size of the female and, except possibly in the case of some Golden Eagles, this is not a reliable field character.

The White-tailed Eagle *Haliaeetus albicilla* is the biggest and heaviest of all the European eagles; its size almost that of a Griffon Vulture. A breeding bird of lakes and coasts, it is rarely found far from water and, like the Imperial and Steppe, soars and glides on flat wings. The Bonelli's Eagle *Hieraaetus fasciatus*, like the *Aquilas*, exhibits a complicated range of plumages and these, too, are linked with age. The Booted Eagle *H. pennatus* has two distinct phases—a light and dark phase which might cause the observer to think they are separate species. The light phase Booted Eagle, Bonelli's Eagle, Short-toed Eagle *Circaetus gallicus* and Osprey *Pandion haliaetus* all show whitish underparts in most of their plumages and although they all glide and soar on more or less flattish wings confusion is really only likely to occur between the last two. As emphasised previously, the wing positions of nearly all birds of prey may alter in relation to wind strength and weather but the normal head-on profiles of this group are shown in Fig. 13.

Unless otherwise stated, descriptions in this chapter refer to the European races, though the problems of subspecies are not usually relevant to field identification. Two taxonomic matters must be mentioned, however. First, the Imperial Eagle has two races in Europe, nominate *heliaca* in the south-east and *adalberti* in Spain, and these are separable in the field. Second, the Steppe Eagle is often considered conspecific with the Tawny Eagle *A. rapax* of Africa and India, but it is in fact rather different and we prefer to treat it as a full species.

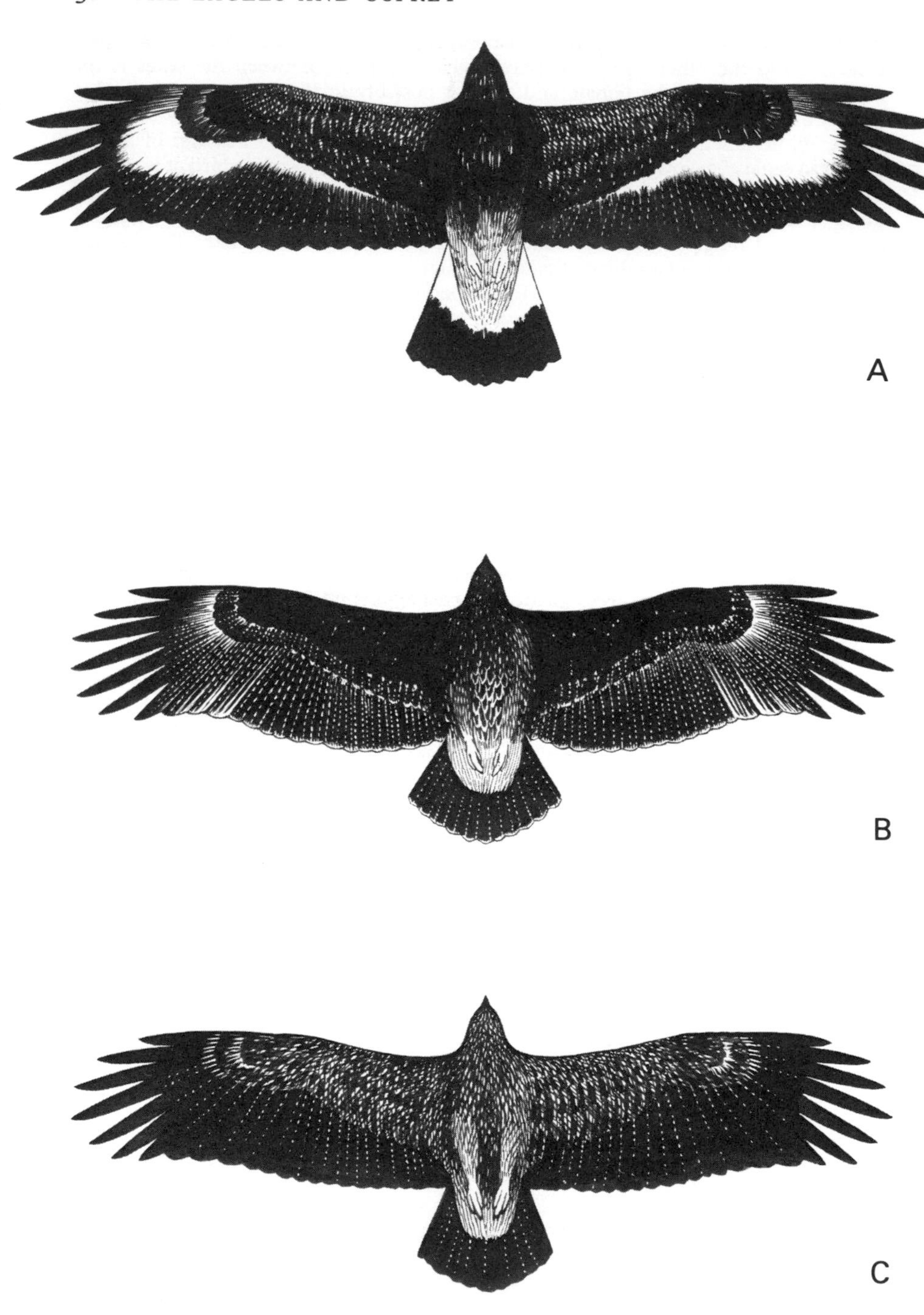

Fig. 14. Typical undersides of juvenile Golden, Spotted and Lesser Spotted Eagles.

Fig. 14A. **Golden Eagle** *Aquila chrysaetos*. Juvenile shown. Large size and long, broad wings with bulging secondaries and narrower hand; fairly long, full, slightly and evenly rounded tail; head and neck protrude fairly noticeable, but not as much as Imperial. Juvenile dark brown with white patches on both surfaces of wings and tail; full adult entirely dark brown except for golden crown and nape, yellowish-brown patch on upperwing-coverts and dark-barred greyish flight-feathers, but younger adults often show immature markings which take many years to disappear. Only European eagle that soars with wings invariably in shallow V. Largely resident in mountain regions of Scotland, Fenno-Scandia and Baltic States across Russia to Urals, and of south and central Europe from Iberia to Carpathians and Balkans, thence to Turkey, Caucasus and Middle East, also larger Mediterranean islands and north-west Africa; partly migratory in Scandinavia and some dispersal in eastern Europe and Middle East in winter.

Fig. 14B. **Spotted Eagle** *Aquila clanga*. Juvenile shown. Smaller than Golden, this and Lesser Spotted being the smallest of the group; relatively short tail and small head not protruding very far. Juvenile with conspicuous rows of spots forming white bars on upper-wing-coverts, whitish uppertail-coverts and whitish patch at base of primaries above; but adult entirely dark brown. Wings invariably held bowed down when gliding. Summer visitor (but generally rare) to damp woods often adjacent to marshland in eastern Europe from south Finland south to Romania and east across Russia; short-distance migrant, but very few on passage at Bosphorus and other traditional migration places; more often seen in winter (but still very scarce) in marshy areas of Turkey and Middle East, also Greece and south France.

See reference to underwing pattern in identification note for this species, page 49.

Fig. 14C. **Lesser Spotted Eagle** *Aquila pomarina*. Juvenile shown. Slightly smaller and better proportioned than Spotted; wings slightly narrower and tail slightly longer. Immature warm chocolate-brown on body and underwing-coverts, contrasting in good light with darker flight-feathers and tail, while above it has rows of white spots on wing-coverts; adult paler brown where immature chocolate-brown and still retains small pale patch at base of primaries and indistinct light U on uppertail-coverts. Soars and glides on slightly bowed wings (less so than Spotted). Summer visitor (far commoner than Spotted) to forest areas in east Europe from north Germany and Baltic States south to Balkans and east across Turkey to Caucasus; passage south-east Europe, Turkey and Middle East, April and end August to late October, but mainly second half September.

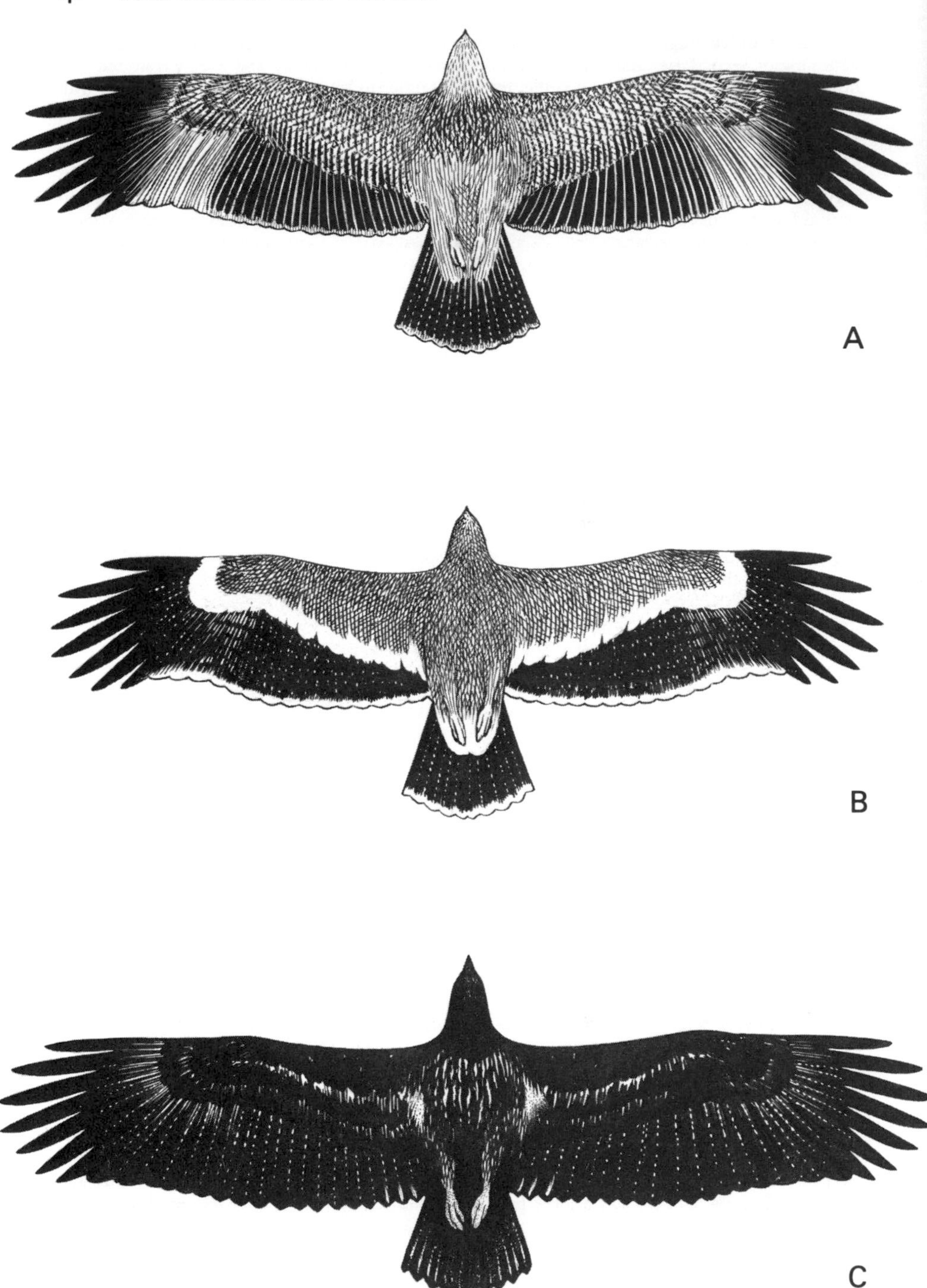

Fig. 15. Typical undersides of juvenile Imperial, Steppe and White-tailed Eagles.

Fig. 15A. **Imperial Eagle** *Aquila heliaca*. Juvenile shown. Large size similar to Golden. Wings and tail fairly long, latter almost width of wings and usually square-cut or slightly rounded at end; head protudes well. Juvenile below has streaked yellowish-brown body and underwing-coverts, and dark flight-feathers and tail with noticeable pale patch on primaries and faint white trailing edge to wings and tail in fresh plumage, while above it is streaked yellowish-brown, paler on lower back, with dark flight-feathers, pale patch at base of primaries and yellowish bar along edge of greater wing-coverts; adult blackish brown with creamy-white crown to hind-neck, white 'braces' and paler basal two-thirds of tail, while Spanish race *adalberti* also has white leading edge to upperwing-coverts. Soars and glides on flattish wings (13B). Spanish race (very rare) largely resident in south Spain and perhaps north Morocco; nominate race resident and summer visitor to plains of south-east Europe and Asia Minor from Hungary to Yugoslavia through Greece, Bulgaria, Romania, south Russia and Turkey; passage at Bosphorus (small numbers) chiefly second half September and first half October, and a few winter on plains and marshes in extreme south-east Europe and Turkey.

Fig. 15B. **Steppe Eagle** *Aquila nipalensis*. Juvenile shown. Size and proportions similar to Imperial. Juvenile like bright version of juvenile Imperial with upperparts, underparts and most wing-coverts unstreaked warm medium-brown, usually brownish lower back and rump, and contrasting whitish uppertail-coverts; greater upperwing-coverts black with white bar along tips, flight-feathers blackish-brown with white band along rear edge and whitish patch at base of primaries. Underside of flight-feathers also dark (but at closer ranges usually appearing distinctly barred) with broad white bands through centre of wing and white trailing-edge; tail dark with white tip. Adult dark grey-brown inclining to almost blackish-brown on underparts (especially on body, carpel patches and wing-tips), often with small rusty-yellowish patch on nape, and usually strongly barred tail and flight-feathers, normally with broad dark band on trailing edge. Soars and glides on flat or nearly flat wings (13B), but characteristically sluggish, more often perched on ground, roadside hummock or telegraph pole than other *Aquila*. Some resident on steppes of Ukraine and other parts of south Russia, and odd pairs may still breed in Romania; others move south in autumn through Caucasus to Turkey, Middle East and beyond; occasionally seen in winter in south-east Europe; closely related Tawny Eagle *Aquila rapax* resident in north Africa.

Fig. 15C. **White-tailed Eagle** *Haliaeetus albicilla*. Juvenile shown. Largest of the group, with broad, rather square-ended and deeply fingered wings; tail very short and wedge-shaped in adult but slightly longer and less wedge-shaped in juvenile; head and neck protrude well in front of wings and powerful bill can be seen at some distance. Juvenile entirely dark brown but for white or yellowish-brown variegations on breast, pale patch on axillaries and diffuse pale bar on median underwing-coverts and rusty-yellow area in centre of upperwing-coverts. Transitional plumages may be rather mottled. Adult grey-brown with white tail and creamy-white head and neck. Soars and glides on flat wings (13B), will sit motionless for hours on ground near water's edge. Largely resident by rocky coasts or inland lakes in Iceland, Fenno-Scandia, Baltic States, Poland and Germany south to Balkans and Turkey; some birds wander in winter in eastern half of Europe, west to Netherlands and occasionally farther west.

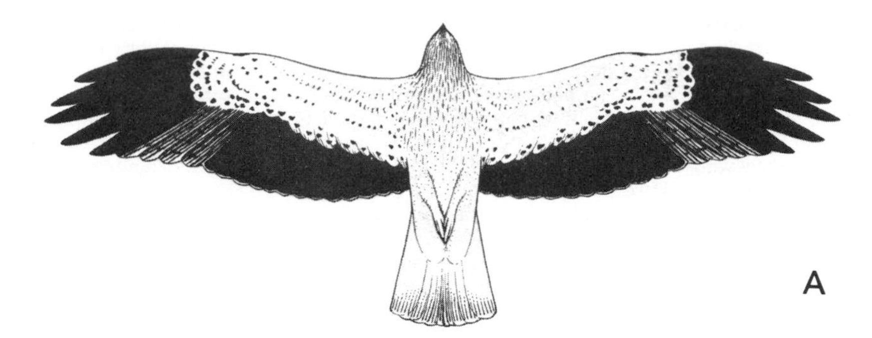

Fig. 16A. **Booted Eagle** *Hieraaetus pennatus*. Light phase. Size similar to Buzzard but shape different with proportionally longer and square-cut tail. Black and white colouration below in light phase eliminates all other European species except larger Egyptian Vulture *Neophron percnopterus*, which has entirely different shape and colouration on upperparts. Watch also for confusion with white forms of Buzzards and Honey Buzzard with densely dark-barred flight-feathers. (For less common dark phase see pages 68–9.) Flight swift and agile with strong deep wing-beats interspersed with glides. Soars and glides on slightly bowed kite-like wings. Iberia, north-central France, Balkans, south Russia, Turkey and north-west Africa, April–October; on passage Mediterranean region from Spain to Middle East, mainly April–May and September.

Fig. 16B. **Bonelli's Eagle** *Hieraaetus fasciatus*. Adult shown. Medium-sized eagle: appears fairly broad in wing and long in body and tail; head small and well protruding; wings held straight out from body when soaring, often with front and rear edges seeming nearly parallel, and shape can be strongly reminiscent of Honey Buzzard *Pernis aprivorus*, especially at distance; tail fairly long and sometimes looking very narrow in contrast to fairly broad wings. Much variation in plumage related to age: adult distinctively black, grey and white, blackest round wing edges, on underwing coverts (which contrast well with pale flight feathers and white body) and at end of tail in broad band; juvenile, immature and sub-adult show various combinations of brown, pale brown and pinkish-rufous (pages 70–73 for details). Soars on wings held level or slightly bowed. Mainly resident (but generally scarce) Iberia, north-west Africa, France, most larger Mediterranean Islands, Greece, Turkey and Middle East.

Fig. 16. Underside of light phase Booted Eagle and adult Bonelli's Eagle.

Fig. 17A. **Short-toed Eagle** *Circaetus gallicus*. Large size, long and relatively broad wings and very pale appearance below are best guide; head broad and thick-set; medium-long tail rather square-cut. Amount of patterning variable: occasional individuals appear almost wholly white below, with very few dark markings, while others show considerable dark patterning on wings and body; small minority lack the dark brown hood seen here or show only traces of it in pale buff-brown, but though variable it is typical of majority. Soars on flat wings and hovers when hunting. Most of Iberia, south and west France, Italy and east Europe north to Baltic States and across Russia to Urals, also Turkey, Middle East and north-west Africa, April–October; passage Mediterranean area, mainly April and September–October.

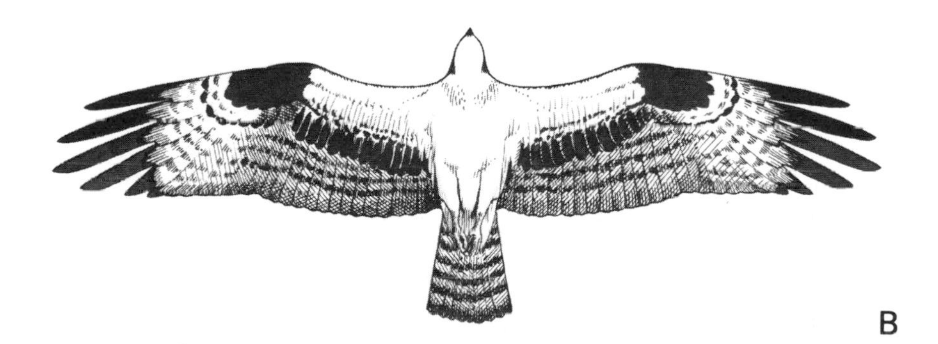

Fig. 17B. **Osprey** *Pandion haliaetus*. Long, slightly angled wings, rather short tail and smallish, protruding head; fairly narrow hand with only five emarginated primaries. Large black carpal patches contrast strongly with white body, white underwing-coverts and generally pale appearance of underwing; buff-brown pectoral band variable in extent; unique black and white head pattern. Soars and, particularly, glides on sharply bowed wings like larger gulls *Larus* spp. with which it can often be confused at distance. Fenno-Scandia and much of east Europe south to Black Sea and Caucasus, also isolated populations in Scotland, south Iberia, north Africa and some larger Mediterranean islands, April–October; passage most of Europe, April and August–October; a few winter around Mediterranean, but most farther south.

Fig. 17. Typical undersides of Short-toed Eagle and Osprey.

Golden Eagle

Aquila chrysaetos
(plates 16–18)

SILHOUETTE Wing-span 204–220 cm. Large eagle, though often looking surprisingly lightly proportioned. Head noticeably protruding, but not so far as that of Imperial Eagle. Long wings usually show distinct S-curve to trailing edge as a result of relatively short inner primaries and innermost secondaries, but wear and moult can make wings look almost parallel-edged; when soaring wings can look oval with bulging secondaries between narrower hand and pinched-in effect at body. Evenly-rounded tail longer and more ample than that of other *Aquila* eagles, length almost corresponding to width of wing. In head-on profile when soaring or, to a lesser extent when gliding, wings raised in shallow V.

FLIGHT Soaring and gliding are most frequent modes of flight. In soaring, wings are raised and pressed slightly forward; in gliding they are less noticeably raised. Active flight usually consists of six to seven deep and powerful wing-beats followed by glide of 1–2 seconds. Most elegant flier of the large eagles and has perfect control even in a very strong wind.

IDENTIFICATION Adult most likely to be confused with adult Imperial Eagle which has relatively long bicoloured tail and very pale crown to hind-neck, but blacker Imperial soars on flattish wings, has slightly more protruding head, less rounded end to closed tail and generally more parallel-edged wings than Golden. Adult Golden normally has greyish flight-feathers above (barred darker and broadly terminated blackish-brown), being uniform blackish-brown in adult Imperial, which at close range often shows diagnostic white 'braces' on scapulars. Distinguished from other *Aquila* eagles by graceful flight, silhouette and raised wing position when soaring; neck and head also protrudes noticeably more than in the two spotted eagles. Juvenile unlikely to be confused with any other raptor, except perhaps Rough-legged Buzzard *Buteo lagopus* in poor conditions because of its similar tail pattern and pale patch on primaries above. It should be stressed that a Golden Eagle often appears smaller than it really is and the observer can frequently be confused by its size unless another species is near for comparison.

Fig. 18. **Golden Eagle** *Aquila chrysaetos*. Juvenile and adult from below. The juvenile (18A) is very distinctive, the underparts being dark brown with conspicuous white patches on the inner primaries and outer secondaries and on the basal two-thirds of the tail. Though these features are visible from both below and above, the wing-patches are usually less extensive on the upperside, whereas on the underside they may almost reach the body as a white point. Individual variation is fairly considerable, as is indicated by the photographs (plates 16–18). The immature gradually loses the white patches as it progresses towards adult plumage (18B), which is attained after about six years. Adult has flight-feathers paler brownish-grey (with variable dark barring and broad dark trailing-edge visible at close range) in contrast with darker warm brown underwing-coverts and body; same applies to tail-feathers where the basal two-thirds are more or less pale and greyish with 2–3 rows of dark bars and the terminal third blackish-brown. In Fenno-Scandia old birds may have warm ochreous-brown breast and median underwing-coverts.

Fig. 19. **Golden Eagle** *Aquila chrysaetos*. Juvenile and adult from above. Juvenile (A) is dark brown with white patches on inner primaries and on basal two-thirds of tail corresponding to those on underside, though wing-patches are usually smaller above and rarely extend to secondaries; again, however, individual variation is considerable (cf. plate 17). At close range yellowish-gold of crown and nape is visible and remains through all ages. Immature (not illustrated) shows golden-yellowish patch across most median and some inner greater upper-wing-coverts, a feature that remains more or less conspicuous throughout all subsequent plumages and is often visible at great distance. Adult (B) is not as dark as young bird, being warmer brown. Basal two-thirds of tail is greyish (barred darker) with dark terminal band; flight-feathers are as below, being brownish-grey barred darker, with a usually distinct broad blackish-brown band along trailing-edge visible only at close range. Some adults may show remains of immature-like whitish markings on base of primaries and tail. Variations in different plumages may be great, thus making age identification in field sometimes dubious.

Spotted Eagle
Aquila clanga
(plates 19–21)

Three Lesser Spotted Eagles and a Spotted Eagle (centre)

SILHOUETTE Wing-span 155–182 cm. Medium-sized, usually compact-looking eagle, distinctly smaller than Golden. As with other *Aquila*, silhouette varies with moult and wear, which affect width of wings and shape and length of tail. Wings moderately long and relatively broad (but some broad-winged birds look relatively short-winged), being quite parallel-edged in some but in many with a slight curve to trailing-edge, caused by slightly bulging secondaries. Generally, wing-tip is more deeply fingered, slightly more square-cut than Lesser Spotted, though wear, moult and individual variation make difference of dubious value. Fresh tail is as long and evenly rounded as Lesser Spotted ($\frac{2}{3}$–$\frac{3}{4}$ width of wings) but may look shorter in comparison to broader wings. In some rather broad-winged birds, a worn tail looks surprisingly short, giving bird an appearance not unlike a small White-tailed Eagle. Head and neck protrude well as in Lesser Spotted but less than in larger *Aquila*. In broad-winged birds head and neck may look relatively small in comparison to width of wings. It should be stressed that much worn and/or heavily moulting birds are difficult to distinguish from Lesser Spotted in outline.

FLIGHT When soaring, wings are carried almost level, sometimes with slightly down-curved hands, and are usually pressed slightly forward. When gliding fast (e.g. on migration), wings are bowed with hand clearly drooping as Lesser Spotted. In slow gliding (foraging) flight, wings may look more drooped than in Lesser Spotted and wing position of a bird flying away from the observer may even recall that of a Heron *Ardea cinerea*. In active flight, wing-beats are usually fast and flat, like Lesser Spotted, but sometimes more laboured (e.g. when foraging) and then beats are heavier with pronounced deep down-strokes, giving bird a characteristic jizz.

IDENTIFICATION Confusion is most likely with Lesser Spotted and Steppe Eagle and great care is needed between these species. Moreover, pale variants of Spotted, so-called *fulvescens* variety, are difficult to distinguish from similar, but very rare form of Lesser Spotted, from pale variants of Steppe, from young pale Tawny Eagles *Aquila rapax* (not found in Europe, though escapes occur) and even from some young Imperial Eagles. For distinction from Steppe and Imperial, see those species. Distinguished from Golden by smaller size, more compact build, shorter, uniformly-coloured tail, less protruding head, heavier flight and entirely different wing position when soaring. In comparison to Lesser Spotted, the Spotted has on average slightly broader wings and, in comparison to these, slightly shorter tail; in both species wings may be parallel-edged or have curved trailing-edge. Though Spotted with the broadest wings, shortest tail and heaviest, most laboured flight differs quite markedly from Lesser Spotted, it is advisable always to look for the more reliable plumage characters, even for the experienced eye.

Normally, underwing-coverts of Spotted are dark blackish-brown (particularly in younger birds) in contrast with slightly to somewhat paler flight-feathers (Fig. 21A–C).

In juvenile Lesser Spotted underwing-coverts are warm to darkish brown, slightly paler than dark grey flight-feathers, though some young Lesser Spotted show an almost uniform underwing without contrast. In immature and adult Lesser Spotted, underwing-coverts are clearly paler than flight-feathers. The following exceptions, however, should be noted: in a few Spotted, flight-feathers below are very dark, almost blackish-grey and underwing looks uniform; colour of underwing-coverts, blackish-brown in Spotted, warm medium to darkish-brown in Lesser Spotted, is then diagnostic. In a few other Spotted Eagles blackish-brown underwing-coverts are mixed with pale blotchings, giving effect of slightly paler coverts than flight-feathers (Fig. 21C): at close range blackish-brown colour rules out Lesser Spotted. Colour of body below, usually blackish-brown in Spotted, warm darkish-brown (juvenile) or paler (adult) in Lesser Spotted (i.e. same as underwing-coverts) also helps identification.

From above, extent of white spotting in juvenile Spotted is often useful for identification. In well-marked, freshly-plumaged birds spotting on greater and median coverts is so broad that they obscure basic blackish-brown colour and wing bars are always much wider than in any juvenile Lesser Spotted (see Fig. 20A). Whitish spotting on lesser coverts can also be very pronounced. In less well-marked birds (more normal) amount and size of whitish spots does not differ significantly from those of young Lesser Spotted, but in these birds upperwing-coverts look blackish-brown at distance with very little or no contrast with blackish-brown flight-feathers, while young (and old) Lesser Spotted has rufous-brown or pale brown upperwing-coverts in contrast to blackish-brown flight-feathers. Beware, however, of bleaching of upperwing-coverts in some adult Spotted, which can give a contrast like Lesser Spotted; in these cases leading-edge of arm (foremost lesser coverts) always remains blackish-brown.

Spotted and Lesser Spotted each show pale patch at base of primaries above. In juvenile Spotted patch stands out as thin whitish shaft-streaks on otherwise dark primaries; in juvenile Lesser Spotted primary webs are paler, thus whitish shaft-streaks are inconspicuous and pale patch looks like yellowish or whitish spot concentrated on base of inner primaries. On average, juvenile Spotted has larger number of whitish spots at tips of primary-coverts and whitish on a larger number of primary shafts than in young Lesser Spotted. Thus pale patch tends to be larger in Spotted, extending more towards leading-edge of wing. However, as it is made up of streaks, it tends to 'disappear' at distance, whereas in young Lesser Spotted it remains as distinct patch. In older birds whitish-coloured shaft-streaks are much reduced or absent, while in adult Lesser Spotted there is at least a pale brownish-yellow patch at base of inner primaries, visible even at distance.

At all ages Spotted lacks rusty-yellow, well-defined nape spot of juvenile and a few immature Lesser Spotted (difficult to see in flight). Spotted has darker head than Lesser Spotted, and yellow cere and gape-flanges stand out conspicuously. Occasionally, bleached adult Spotted may show pale greyish-brown head-hind-neck.

In Spotted there is tendency to show more pure white on all uppertail-coverts and a sometimes larger area of whitish-grey at bases of outer primaries below (seen as a whitish crescent spot) than in Lesser Spotted.

In *fulvescens* variety there is great contrast between very pale wing-coverts and dark flight-feathers, both above and below, and it is advisable to look for other characters such as primary-patch, flight and silhouette.

Fig. 20. Spotted Eagles. Juvenile, immature, adult and (opposite) 'fulvescens' from above.

Fig. 20. **Spotted Eagle** *Aquila clanga*. From above and, overleaf, from below. Juvenile (A), immature (B), adult (C) and 'fulvescens' (D) juvenile (E), adult (F) and 'fulvescens' (G). Juvenile above is a blackish-brown bird with purple gloss in fresh plumage and variable amount of whitish markings. Well-marked, fresh plumaged juvenile (A) has whitish tips to secondaries and innermost primaries, tail-feathers, upperwing-coverts, longest scapulars, mantle and, most frequently, whitish spotting on lower back (sometimes appearing as a whitish patch). Uppertail-coverts are white and there is pale patch on primaries formed of white shaft-streaks. No young Lesser Spotted would ever show so much white on upperwing-coverts. In less-well-marked juveniles white spots are much smaller and pattern does not differ so greatly from that of young Lesser Spotted, but, of course, in Spotted ground colour is blackish-brown with little or no contrast to flight-feathers. Whitish trailing-edge to wings and tail may wear off and in transitional plumages all white markings are reduced and normal, worn immatures above (B) look blackish-brown and only pale marking may be one narrow wing-bar on greater-coverts, perhaps some whitish on lower back, white on uppertail-coverts and on primary shafts. Below, normal juvenile (E) has blackish-brown body and underwing-coverts, contrasting more or less, depending often on light conditions with dark grey flight feathers (and greater-coverts), which may be faintly barred; sometimes the three inner primaries may be a little paler than rest of flight-feathers. Whitish crescents around carpal patch (E) are often present; pale undertail-coverts and vent, extending in some onto lower belly; in few the greater secondary coverts faintly pale-tipped, and in fresh plumage juvenile shows whitish trailing-edge to wings and tail, which becomes lost through abrasion.

Adults (C) lack white spotting above and white trailing-edge to wings and tail, though few may show diffuse pale bar on greater coverts. White shaft-streaks on primaries and uppertail-coverts are reduced; pale patch on lower back occasionally present. Thus from above adult looks almost uniform dark brown with little or no contrast between dark brown upperwing-coverts and blackish-brown flight-feathers. However, in some, median and most lesser coverts bleach paler to give Lesser Spotted-like contrast. Below, adults (F) resemble juveniles with same contrast on underwing. Flight-feathers may be all uniform or primaries may be slightly paler than secondaries with inner primaries palest; they may be narrowly barred or completely unbarred. Pale bases to outer primaries and greater primary coverts present or lacking as in juvenile but vent and undertail-coverts usually dark brown.

Some plumages below (not illustrated) may be particularly difficult to identify but these are rare. For example, body is pale brownish in some but underwing-coverts are still blackish-brown. In others body is dark but dark coverts are irregularly mottled yellowish, giving at distance a dark medium-brown appearance in contrast with dark flight-feathers, very like some Lesser Spotted Eagles (and also some Steppe, which see).

Typical 'fulvescens' (D and G) have, in fresh plumage, above very broad creamy edges and tips on brown upperwing-coverts, and forewing looks very pale in contrast to dark flight-feathers.

Creamy-yellow head, neck and upperparts, frequently with darker lower mantle and even paler lower back and rump; white rump. Whitish trailing-edges to wings and tail disappear with wear as do pale edges to coverts, and forewing may look just pale brown. Below 'fulvescens' is uniform creamy-yellow on underwing-coverts and body in strong contrast to dark greater-coverts, flight- and tail-feathers. There is doubt as to whether 'fulvescens' form occurs in adults.

Fig. 20. (cont.). Juvenile, adult, and 'fulvescens' Spotted Eagles from below.

Lesser Spotted Eagle
Aquila pomarina
(plates 22–24)

SILHOUETTE Wing-span 134–159 cm. Distinctly smaller than Golden Eagle. Often compact-looking, though generally with slightly narrower wings than Spotted, giving a longer-tailed appearance in comparison to width of wings. However, wear may cause tail to look shortish. Wings frequently rather parallel-edged but some show slight curve on trailing-edge. Generally, wing-tip a little more rounded and 'fingers' less deep than in Spotted, but differences are slight and inconsistent.

FLIGHT Wing position when soaring and gliding as in Spotted. Generally, Lesser Spotted looks slenderer and less heavy in the air, with shallower and lighter wing-beats than Spotted. (See also Spotted Eagle.)

IDENTIFICATION Easily confused with Spotted and Steppe Eagle; pale or mottled variants are very hard to identify if no size comparison is possible. Distinguished from Golden and White-tailed Eagle in same way as Spotted Eagle (which see). Although distinction from Spotted is given under that species, the following should be noted: below, Lesser never shows any blackish-brown on underwing-coverts and body; whilst, above, contrast between blackish-brown flight-feathers and paler brown median and lesser coverts is usually valid as is shape of pale primary patch.

Distinction from some adult and sub-adult Steppe Eagles requires great care, particularly if the two species are not together. Lesser Spotted is smaller bird (obvious when seen together) with smaller bill and less protruding neck; tail is slightly shorter and, when fresh, evenly rounded (and not slightly wedge-shaped). Wings are comparatively shorter, hand not so long and ample and trailing edge much less frequently as S-curved as it is in many Steppe Eagles; wing-beats are faster, shallower and less heavy. These characters require experience and are often difficult to establish when the two species are not together, and frequently it is combined plumage characters that identifies the eagle. No Lesser Spotted shows broad white band on greater underwing-coverts and broad white trailing-edge to wings and tail of juvenile plumage Steppe. Difficulties arise when, through moult and wear, Steppe loses these features. Since Lesser Spotted has paler wing-coverts than flight-feathers, both above and below, it can be confused only with those Steppes which have same contrasts (i.e. adult Steppes with uniform underwing, or darker coverts than flight-feathers, are easily ruled out). In such cases Lesser Spotted shows medium to darkish brown body in contrast to paler brownish underwing-coverts, contrasting again to dark grey flight-feathers. At close range both species may show narrow whitish line at tips of greater coverts below, but Lesser Spotted nearly always lacks clear-cut white edgings (and, if present, only as a few diffuse whitish edges) shown by Steppe especially on

(*continued p. 56*)

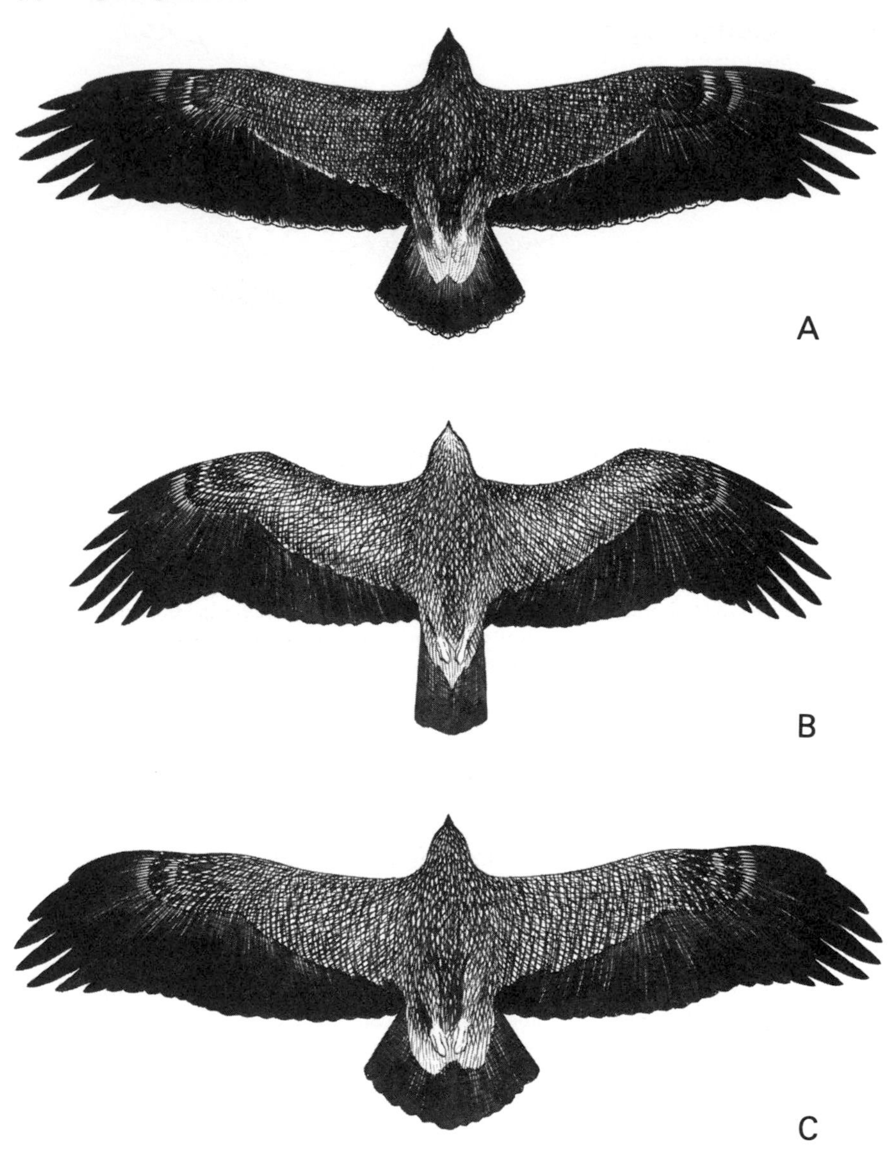

Fig. 21. **Lesser Spotted Eagle** *Aquila pomarina*. Juvenile and two adults from below and, opposite, juvenile and adult from above. Normal juvenile (A) has warm medium- to darkish-brown body and underwing-coverts, contrasting more or less with darker grey-to-blackish-grey flight-feathers (which may be faintly barred). Three inner primaries sometimes very slightly paler than other flight-feathers or all primaries very slightly paler than secondaries. Whitish crescents at base of outer primaries and bases of greater coverts often present. Pale brownish-yellow vent and undertail-coverts (in some even invading lower belly) normal in juvenile. In many, greater secondary coverts are tipped whitish to form thin line down middle of wing. Fresh-plumaged juveniles (A) show whitish trailing-edge to wings and tail, which are lost through wear. Immatures (illustrated on plates 22 and 24) frequently have irregularly mottled paler brown underwing-coverts in contrast to darker brownish and often irregularly mottled body and, of course, dark flight-feathers; whitish line on greater coverts can be quite conspicuous (and lead to confusion with Steppe Eagle), or absent. Adult B and C has dark flight-

and tail-feathers as juvenile (but lacks whitish trailing-edge). Pale crescents at bases of primaries can be present or lacking as in other plumages, but vent and undertail-coverts usually little paler than medium-brown underparts.

Juvenile from above (D) is warm darkish-brown, slightly paler (warm mid-brown) on median and lesser wing-coverts in contrast to blackish-brown flight-feathers. In fresh plumage there is variable amount of whitish drop-like spotting on greater and median coverts; narrow whitish tips to secondaries and innermost primaries and tip of tail-feathers quickly lost through wear. Uppertail-coverts whitish and frequently there is variable whitish spotting on dark brown lower back, sometimes featuring as whitish patch. Small rusty-yellow, well-defined nape-patch is often difficult to see. There is a small pale area at bases of inner primaries. In immature transitional plumages white spotting becomes less and some wing-coverts become paler brown. Adult (E) lacks conspicuous white spotting and whitish trailing-edge to wings and tail, though a few may show diffuse pale bar on greater coverts at very close range. Pale primary patch is smaller but always present and pale uppertail-coverts are usually less distinct. Back is dark brown grading into medium-brown mantle. Head, median and lesser coverts are pale brown, in some bleached adults inclining to yellow-brown and in clear contrast to blackish-brown flight-feathers. Rusty-yellow pale variants resembling Spotted Eagle 'fulvescens' occur rarely.

greater primary coverts. Lesser Spotted has much less barred flight- and tail-feathers than Steppe and frequently lacks any barring at all; it also lacks often conspicuous dark band on trailing-edge of wings and tail seen in most adult and sub-adult Steppe. In Lesser Spotted, pale primary-patch is generally smaller, not extending so much towards leading and trailing-edge of hand, though this is variable. Again, at close range small bill and less extensive and less conspicuous yellow gape-flanges (visible in photographs) distinguish the most odd-looking Lesser Spotted from Steppe. Rusty-yellow, well-defined nape-patch of young Lesser Spotted can also be seen in Steppe but, while it is lost in adult plumages in Lesser Spotted, it occurs more frequently in Steppe with increasing age. For distinction from Imperial Eagle see that species.

Imperial Eagle

Aquila heliaca

(plates 25–28)

Adult of Spanish race 'adalberti'

SILHOUETTE Wing-span 190–210 cm. Large, heavy and long-winged eagle, near Golden in size. Wings may appear fairly broad in some, particularly younger birds, due to long secondaries and short inner primaries producing an S-curve on trailing edge of wings. In others, largely non-juvenile and adult birds, wings can look narrower and rather parallel-edged. Hand is long and ample, and primary 'fingers' long and deeply cut. Fresh tail is fairly long, nearly width of wings (like Golden Eagle) and almost square-ended when closed. Large bill and long neck protrude more than in other *Aquila* eagles.

FLIGHT When soaring wings usually held level and almost at right-angles to body, sometimes very slightly lifted and pressed forward; longest primary tips are a little more conspicuously lifted than other *Aquilae*. When gliding slowly wings usually level; in faster gliding they are angled with carpal joint pressed forward and hands pointing backwards and slightly lowered. When soaring tail is frequently held closed, looking narrow. Active flight is heavy, though less clumsy than Steppe's (e.g. on migration), consisting of three to ten deep and slightly elastic wing beats, interspersed with glides. Adults appear less heavy, more slender and slightly more agile than juveniles and immatures.

IDENTIFICATION For distinction from Golden and White-tailed Eagle see those species. Imperial is larger, longer-winged and longer-tailed (when fresh) with more protruding head and neck than the two spotted eagles; also, in fresh juvenile plumage has more curved trailing edge to wings. Wing position when soaring and gliding is flatter. Adult Imperial's bicoloured tail (uniform in spotted eagles) and well-demarcated yellowish-white crown to hind-neck are useful distinguishing features. Adult Imperial show same underwing contrast (blackish-brown coverts, paler flight-feathers) as Spotted Eagle; vent and undertail-coverts usually a well-defined rusty-yellowish patch; paler base of undertail, which looks semi-translucent in certain lights, accentuates broad blackish terminal tail-band. Some adult Lesser Spotteds have much paler underwing-coverts than flight-feathers, thus superficially resembling young Imperial, though latter species has distinctly paler innermost primaries below, contrasting with remaining blackish-brown flight-feathers with whitish trailing-edge in fresh plumage. Another useful feature of young Imperial (nominate race) is the distinctly streaked lower neck and upper belly, looking at distance like broad pectoral band in contrast with the very pale hind body. Above, young Imperial can easily be confused with *fulvescens* form of Spotted Eagle, but from below that species lacks breast streaking of young nominate Imperial, as well as lacking pale inner primaries of underwing. Imperial in transitional plumage normally has more pied blackish-brown/rusty-yellow body and underwing-coverts than most variants of spotted eagles and bicoloured tail and pale crown to hind-neck start to show up. Some still show pale inner-primaries below. Normally, the experienced observer should have little difficulty in distinguishing an Imperial from the two spotted eagles and it can often be done on silhouette alone. Confusion with Steppe Eagle is, however, more likely.

Generally, Imperial is more slender with slightly longer, more square-cut tail and with slightly more protruding neck than Steppe Eagle. Individual variation in wear and moult, however, mean these characters cannot be used without studying plumage. Young Imperial lacks broad white band on underwing of young Steppe and can only be confused with those few individuals in which this white band is lost. Young Imperial of nominate race has dark streaked forebody which Steppe lacks. Secondaries below of young Imperial in fresh plumage are blacker with greater contrast to paler innermost primaries than average Steppe which, on the other hand, has generally much more distinctly barred flight-feathers. Young Imperial is also often a considerably paler yellowish bird than young Steppe. Variations, however, do occur in both species and thus emphasis should be placed on Imperial's streaked breast (nominate race) and Steppe's white band on underwing.

From above, young birds of both species are very similar, but at close range young Imperial of nominate race shows more or less streaked upper-wing-coverts, scapulars and mantle, these parts being uniform in Steppe. In young Imperial lower back and rump are much more frequently pale to form large creamy-white patch accentuated by white uppertail-coverts; in young Steppe (which also has white rump) lower back and rump are most frequently uniform medium brown with just a small pale spot at base of lower back; pale primary-patch is generally smaller in young Imperial which does not show such strikingly white primary-shafts as Steppe.

In transitional plumage Imperial's blackish-brown/rusty-yellowish mottled under-wing-coverts and forebody, contrasting sharply with uniform rusty-yellowish lower belly to undertail-coverts, separates it from the most mottled Steppe at close range. In early stages of immature plumage Imperials usually show pale inner primaries below, contrasting with very dark secondaries. However, main points to look for are usually well-defined golden-yellowish crown to hind-neck and bicoloured tail, best seen from above. A few Imperials in transition show diffuse palish band on median/lower lesser

Fig. 22. **Imperial Eagle** *Aquila heliaca*. Juvenile, immature and adult from below and (on page 60) from above. From below, juvenile is brownish-yellow with variable dark streaks on body and wing-coverts, with greyish greater coverts and contrasting darker flight- and tail-feathers; streaking absent or confined to upperbreast in paler, more rusty Spanish race *A. h. adalberti*. In nominate race streaking usually seen as dark pectoral band (see plates 25 and 26) contrasting with pale unstreaked hindbody; in some (A), streaks are thinner and less conspicuous. In fresh plumage, flight-feathers are blackish-brown, with distinctly paler yellowish-grey three inner-most primaries and narrow whitish-yellow bar along trailing-edge of wing and tail, which is lost when worn. From above (D and plate 27) shows blackish-brown flight-feathers and tail, all with pale trailing-edge, and pale three inner primaries which together with pale on outer webs of remaining primaries form pale 'wedge'. (Note: no white primary shafts.) Whitish-yellow wing-bar on edge of dark greater coverts and less conspicuous pale tips to median coverts may also form bar, but much less obvious. Head, mantle, back and forewing-coverts brown with creamy-white tips to feathers, overall impression being of creamy plumage with darker streaks; lower back and rump paler (but lowest rump feathers a little darker) and uppertail-coverts white.

Transitional plumage (B and E and plates 26–27) commences after about three years; in early stages looks very yellowish, but blackish-brown feathers start to appear on lower neck/breast (plate 27) and then on wing-coverts; later, dark feathers dominate and underparts look patchy, though vent-undertail-coverts remain pale (plate 26). Patchy/pied underwing-coverts do not form constant patterns, but lesser coverts often darker earlier than median, thus forming pale bar through centre of underwing (compare with Steppe Eagle). Flight-feathers are blackish-grey and the three pale inner primaries have dark tips and dark barring. Adult-like bicoloured tail pattern shows up (above and below) from early transitional plumage. Above, transitional bird shows reduced primary patch, diffuse pale bar on greater coverts (only visible at close range). Wing-coverts are variably mottled at first being mixture of worn juvenile and new blackish-brown and yellowish feathers. Mantle dark brown, and rusty-yellow crown-hind-neck starts to show up. Frequently there is pale patch on lower back, while rump is normally dark and uppertail-coverts white or brownish with pale tips. Adult (C and F) blackish-brown on underparts, median and lesser underwing-coverts in some contrast to paler dark grey greater coverts and flight-feathers; rusty-yellowish vent-undertail-coverts usually form contrast with very dark belly, and bicoloured tail pattern is not so conspicuous as above. Upperwing is blackish-brown with little or no pale primary-patch, though sometimes paler brownish on parts of wing-coverts (F). Variable amount of white on scapulars, in some this shows as white V surrounding mantle. Lower back and rump uniform or with pale brownish patch on lower back; brownish uppertail-coverts sometimes tipped white. Note characteristic tail pattern with broad blackish sub-terminal band and almost silvery-grey, dark barred basal two-thirds. Spanish race *A. h. adalberti* has conspicuous white leading-edge to coverts and more prominent white scapulars.

wing-coverts; this may lead to confusion with Steppe, though in that species, of course, band is on the greater underwing-coverts. As in juvenile plumage pale primary-patch above is smaller or absent in Imperial. Adult Imperials are blacker than most adult Steppes and have on average darker and much less distinctly barred flight-feathers below which are paler than the underwing-coverts; they have less distinct dark band on trailing-edge of underwing and much more frequently rusty-yellow vent to undertail-coverts (dark brown in most adult Steppes) contrasting with blackish-brown body. Rather pale base of tail accentuates dark band at tip, and yellowish-white crown to hind-neck can be seen at great distance. A few dark adult Steppes show large, rusty-yellow, well-defined patch covering crown and nape but this does not extend to hind-neck. When close, white 'braces' on scapulars immediately distinguish adult Imperial, Spanish *adalberti* form of which also shows conspicuous white leading-edge to upperwing.

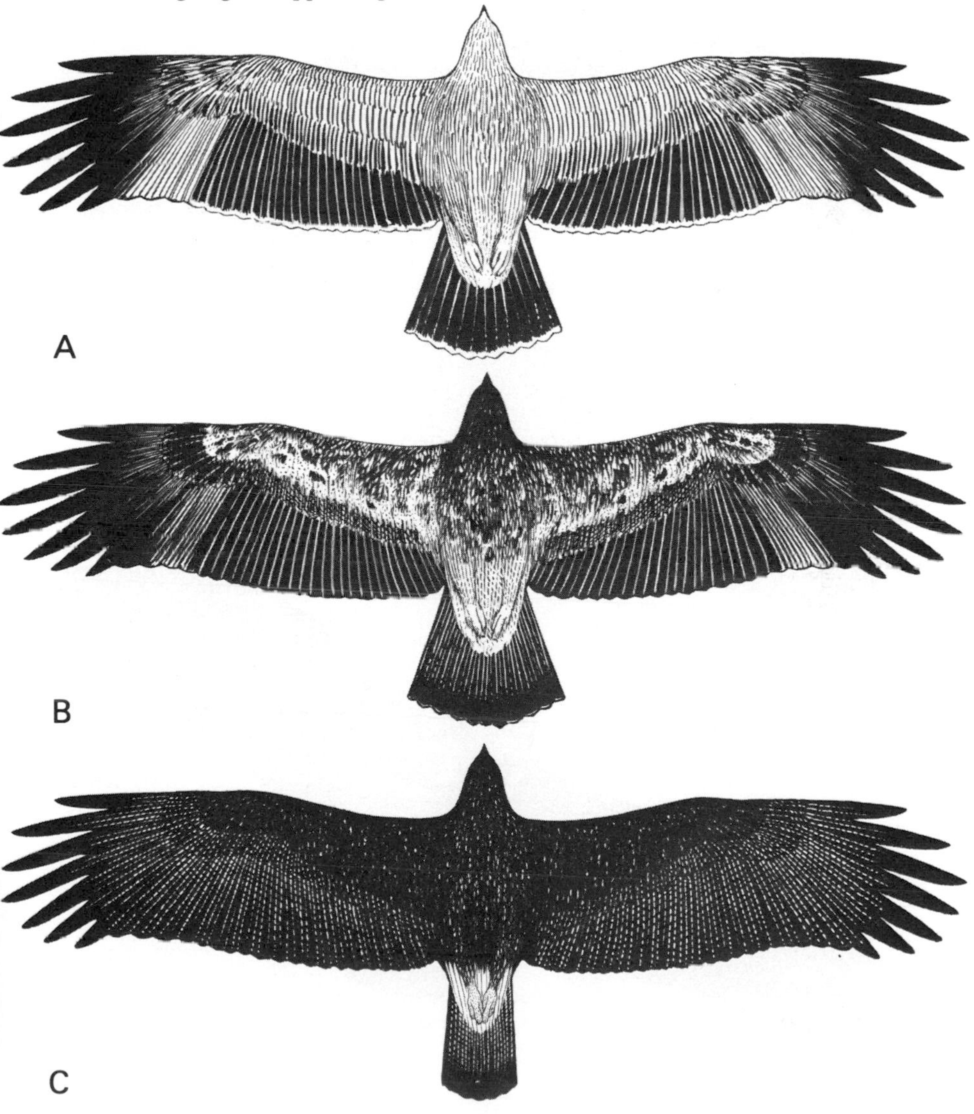

Fig. 22. Imperial Eagle. Juvenile, immature and adult from below (see overleaf for views from above).

Fig. 22. (cont.). Juvenile, immature, and adult Imperials from above.

Steppe Eagle
Aquila nipalensis
(plates 29–32)

SILHOUETTE Wing-span 174–260 cm. Large, heavy and long-winged eagle, eastern race *A. n. nipalensis* being similar in size to Imperial, whilst European subspecies *A. n. orientalis* is smaller. Wings usually appear rather broad with long ample hand and deeply cut 'fingers' and with distinct to slight S-curve to trailing-edge, though many adults show practically parallel-edged wings. Others may even look more narrow-winged and all intermediate shapes occur not least as result of moult and wear. Fresh tail is fairly long, being about ¾ width of wings; when closed is slightly wedge-shaped to rounded, but wear and moult can produce short-tailed appearance, especially in adults. Relatively long bill and neck protrude well in front of wings, almost as much as in Imperial.

FLIGHT When soaring, wings usually held level or slightly bowed with lowered hand (like soaring Spotted Eagle or gliding Red Kite *Milvus milvus*). When gliding slowly, wings typically much more bowed with distinctly drooping hand, at least as much as Spotted. When gliding faster, the kinked wings recall wing position of fast-gliding spotted eagles (slightly lifted arm, lowered hand). Generally wings are not so flat as in Imperial. Active flight is heavy and clumsy; on migration, flight is usually about 3–10 deep and heavy wing-beats, interrupted with glides, but when foraging greater number of wing-beats often involved.

IDENTIFICATION For distinction from Golden, Lesser Spotted and Imperial Eagle see those species. Steppe Eagle is larger, longer-winged and longer-tailed than Spotted with generally more curved trailing-edge to wings (due to shorter inner primaries) and longer, more ample hand, but wear, moult and individual variation make separation on outline alone difficult. Most evident is longer neck and bill of Steppe, though again this can be a difficult point. Younger birds with their pale plumage and broad white band on underwing are unmistakable. Pale immatures without broad white band and pale (sub)adults can be confused with *fulvescens* variety of Spotted but these forms are very rare. Equally difficult is separation of typical adult and sub-adult Steppe from adult Spotted, especially if seen alone in worn plumage or at far range.

At distance many adult Steppe show uniform dark underwing-coverts with slightly contrasting paler flight-feathers, very similar to typical adult Spotted, and they are then virtually inseparable. The often distinctly barred flight- and tail-feathers in adult Steppe, with broad blackish band along their trailing-edges, are lacking in Spotted but these features are visible only at close range. In some adult Steppes, however, dark band to trailing-edge of wings is so pronounced that it can be seen at a fair distance in good light and is then best character from below. Adults with diffusely barred flight- and tail-feathers can be identified at close range by a combination of: often more variegated underwing-coverts, white edgings to some greater coverts (when present), dark brown vent/undertail-coverts (exceptions occur and a few adult Spotted are just as dark) and by facial pattern. Steppe has extensive yellow gape-flanges, reaching rear edge of eye (not more than centre of eye in Spotted), which are separated from pale patch on centre of throat by dark line bordering lower edge of flanges, a pattern not seen in any other *Aquila* and visible up to about 50 m. Most Steppes also show well-defined rusty-yellow nape-patch. Upper sides of flight-

and tail-feathers are often barred, as below, with dark band to trailing-edges (lacking in Spotted).

Pale primary-patch above, formed by whitish shafts, is a useful character even at a distance; despite variation it is larger and paler than Spotted, sometimes covering most of hand; the three inner primaries at least are always paler and show as an ill-defined 'wedge' on the upperwing. In adult Spotted, whitish shaft-streaks may stand out on dark hand (never so in adult Steppe) but streaks can be almost lacking and some adult Spotted show virtually no pale primary-patch at distance. Rare pale brownish-yellow (sub)adult Steppe may look similar to *fulvescens*-types of Spotted (also rare), both having much paler coverts than flight-feathers above and below. Barring on flight- and tail-feathers below and pale primary patch above distinguish such Steppes; when present, clear-cut white edging to some greater coverts below helps separation at close range. Large size of bill is also a useful character at close range but facial pattern is obscured by pale colouration on head.

Separation from Imperial is mentioned under that species. It should be stressed that they are often inseparable on silhouette. Steppe's white band on underwing (present on some into adult plumage) always distinguishes it from Imperial. Steppe has larger primary-patch above, and on average more barred flight- and tail-feathers most pronounced in adult plumage. Most adult Steppes have underwing-coverts of same shade as flight-feathers; in some, coverts are paler and confusion here is most likely with Lesser Spotted, (which see); in a few, coverts are darker and thus similar to adult Imperial but earlier comments should prevent confusion. Most show characteristic dark wing-tip, perhaps dark carpal-patches and dark body. (See Plates 31 and 32).

Fig. 23. **Steppe Eagle** *A. nipalensis*. Juvenile, immatures and adult from below. Most juveniles (A) are medium brown to pale clay-brown on body, lesser and median wing-coverts with white greater coverts forming broad band on underwing. Vent and undertail-coverts whitish-yellow and centre of throat also paler, forming a diffuse patch. Flight- and tail-feathers dark brown,

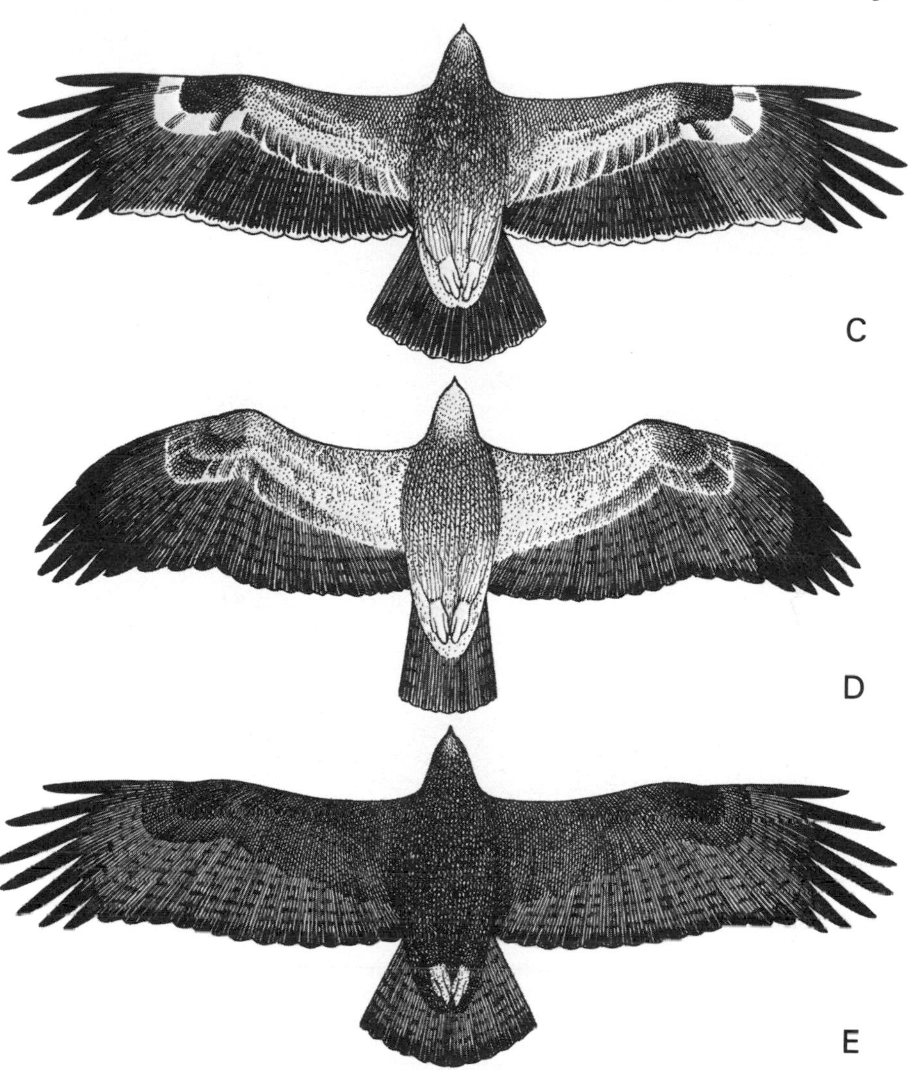

variably barred with broad yellowish-white band on their trailing-edge which is lost through abrasion. In most, inner primaries appear paler than other flight-feathers. Transitional plumages (B and C) are highly variable. Typical second autumn bird shown in B, with heavily worn secondaries and tail, and white underwing band already becoming sullied and mottled with new, dark feathers. Later, body and carpal-patches darken faster than underwing-coverts, which may be irregularly mottled pale and dark brown, especially in lighter forms (C). White greater covert band usually remains throughout the transitional plumage, but in end may be just formed by white edgings and tips to greater coverts, though most of the primary coverts may be white. During transitional plumages barring on flight- and tail-feathers usually becomes more conspicuous and some sub-adults show dark band to trailing-edge of wings (D). Adults (E) have most barred flight- and tail-feathers, usually with broad dark band to trailing-edge. Variation in barring can render flight feathers dark or palish at distance. Most adults have dark grey greater coverts but many still show a few whiter edges and tips. Body and undertail-coverts usually dark brown; in some underwing-coverts are just as dark as body and contrast slightly with paler flight-feathers; in others they are paler greyish-brown, similar to flight-feathers and contrast with darker brown body; in a few, coverts are paler brown in contrast to darker body and flight-feathers. Nearly all have dark carpal and wing tips (E) and pale patch on centre of throat visible at close range.

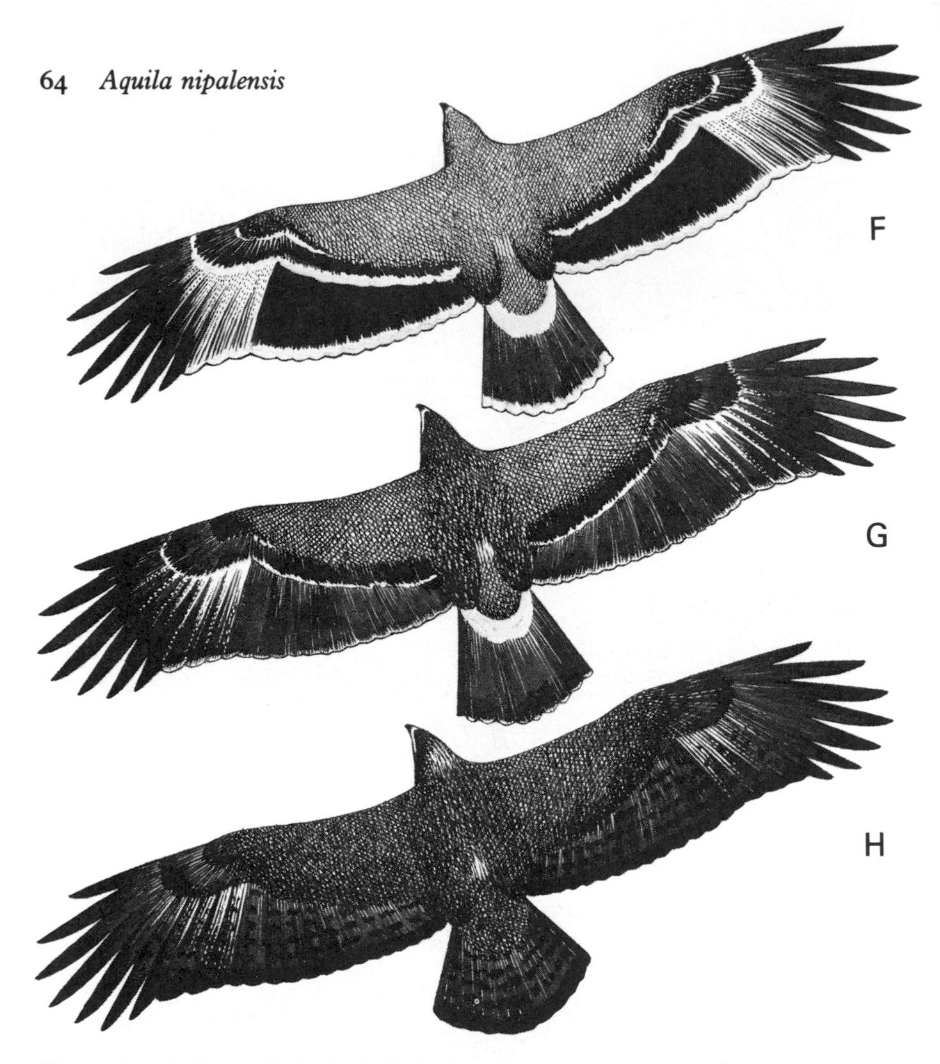

Fig. 23. (cont.). **Steppe Eagle** *A. nipalensis*. Juvenile, immature and adult from above. Juvenile (F) in fresh plumage is strikingly patterned with broad white bars on tips of blackish-brown greater coverts, secondaries and tail; large pale primary patch which is variable in size and paleness; start of second wing-bar on carpal joint formed by white tips to outer median coverts—again a variable feature. Rest of wing-coverts and upper parts typically medium or pale brown (but colour may range from yellowish to dark brown) with whitish area on lower back which is usually small and obscure (F), occasionally large (plate 29), but rarely extending onto rump which is dark in contrast to white uppertail-coverts. Variable pale patch on lower back is present in all ages. White trailing-edge wears almost off wings and completely off tail and coverts bar may also disappear.

In transitional plumages upper parts gradually become darker brownish (G) but there is much variation and sub-adults can show rather pale brownish-grey or yellowish-brown wing-coverts, usually with darker body. Barring on tail and flight-feathers becomes more distinct with increasing age. Most immatures/sub-adults have white uppertail-coverts with dark smudging in older birds. Occasionally immatures show rusty-yellow patch on nape.

Adults are variable in their barring on flight- and tail-feathers (H and plates 31–32). In more barred specimens (dark band at trailing-edge of wings and tail characteristic at close range, but in some, barring poorly developed and flight- and tail-feathers look uniform dark brown (wear can also cause this feature to disappear). Pale primary-patch is always present but variable in size and paleness, though usually distinct and even appearing translucent. Barring on this patch characteristic but often poorly developed. Upperwing-coverts usually grey-brown but sometimes even yellowish-brown, with head and back darker, except for well-defined rusty-yellow patch on rear crown/nape, but this is often diffuse or lacking.

White-tailed Eagle
Haliaeetus albicilla
(plates 33, 34)

SILHOUETTE Wing-span 200–240 cm. Large and heavy eagle, size as Golden or bigger (compare Figs. 14A and 15C). Adult bird has long broad wings with almost parallel edges, short wedge-shaped tail, approximately half or two-thirds of width of wings and massive bill and long neck protruding as much in front of wings as tail does behind. Juveniles have more S-curved trailing edge to wings, due to slightly longer secondaries whose tips are characteristically jagged. Tail of juvenile averages somewhat longer and less wedge-shaped than adult, sometimes only slightly rounded when closed. In head-on profile when soaring or gliding, wings level and flat, though sometimes a bend is visible at carpal joint with slightly lowered hand when gliding.

FLIGHT In soaring and gliding, wings are held flat and almost at right angles to the body, though when soaring in thermals they may become very slightly raised, but never so much that confusion could arise with Golden Eagle. In active flight, large size and broad wings may give heavy and clumsy appearance, but wing beats are often fairly rapid and shallow, and interrupted by glides on level or slightly lowered wings. When flying purposefully from one point to another, however, periods of gliding are often reduced in length and frequency.

IDENTIFICATION Adult cannot easily be confused but juveniles are sometimes misidentified as *Aquila* species despite looking very heavy. Wing position and wing and tail-pattern distinguish it from Golden (which see) but inexperienced observer can confuse it with other *Aquilae*, even the much smaller Spotted. All *Aquilae* can show wedge-shaped tail when worn, and White-tailed can glide on drooped wings like Spotted/Steppe Eagle or on flatter wings like Imperial. Young White-tailed, however, is heavier in the air, has very long 'fingers' on broad wings and long, slender neck and huge bill should rule out other species. At close range the following may be noted: whitish axillary patch, palish bar on median underwing-coverts (sometimes seen in Imperial and often in Steppe), whitish on inner secondaries, whitish-grey centres to most tail-feathers (best seen when tail spread) and unbarred flight-feathers; also never shows clear-cut white uppertail-coverts or pale primary patch above as can be seen in young *Aquilae*. White-tailed in transitional plumage can look superficially like an immature Imperial, but best distinguished by white in short wedge-shaped tail and much heavier appearance. Distinguished from Griffon Vulture *Gyps fulvus* and Black Vulture *Aegypius monachus* by much more protruding neck; Griffon is usually much paler on wing-coverts and soars on raised wings, while Black Vulture is blacker on wing-coverts; both lack whitish axillary patch and have no white in tail.

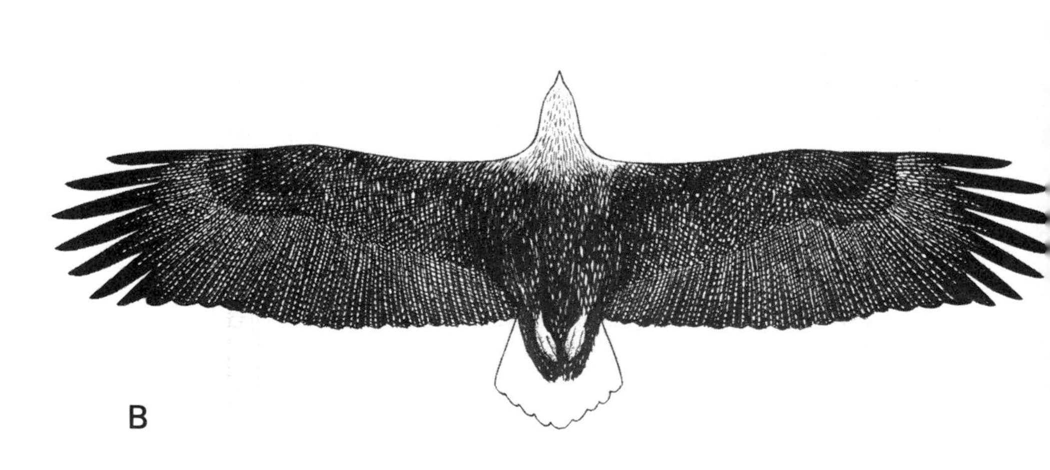

Fig. 24. **White-tailed Eagle** *Haliaeetus albicilla*. Juvenile and adult from below. Blackish-brown juvenile (A) has variably rusty-yellow streaked breast (head to neck often darker), usually distinct whitish axillary patch and often a palish bar at base of median coverts. Whitish-grey centres to most tail-feathers visible on spread tail which otherwise looks all dark at distance. Unbarred flight-feathers (all ages) are dark greyish, slightly paler than blackish-brown on most wing-coverts; wing-tip also darker; few whitish markings visible on innermost secondaries and greater coverts. Immatures are variably mottled rusty-yellowish and dark brown on wing-coverts and body, and whitish axillary patch and pale bar on median coverts present in many birds well into sub-adult plumage. Amount of white in tail increases and, in some, tail pattern can approach that of young Golden Eagle. Frequently, lower underparts mottled dark and light whilst head to upperbreast all dark. Adult (B) is dark grey-brown below, paler brownish or yellowish-brown on upper-breast grading into paler head, even creamy-white in older birds; tail is completely white.

Fig. 25. **White-tailed Eagle** *Haliaeetus albicilla*. Juvenile and adult from above. Juvenile (A) is blackish-brown above with pale rusty-brown panel on wing-coverts, variable pale rusty-brown mottling on back, slightly paler diffuse patch on primaries and whitish on innermost secondaries and coverts; dark brown tail shows greyish-white centres to most feathers when spread; white loral patch contrasts with blackish-brown head and dark grey cere. Immatures (not shown) are variably mottled and variegated on wing-coverts, back and mantle. White in tail increases with age and, though variable, is frequently dirty white with irregular-shaped dark terminal band. Adult (B) has yellowish-grey-brown upperwing-coverts, white tail and, particularly in old birds, creamy- or silvery-white head. Dark grey flight-feathers show diffuse palish patch on primaries. Bill all yellow.

Booted Eagle
Hieraaetus pennatus
(plates 35, 36)

Dark phase Booted Eagle (left) and dark Marsh Harrier

SILHOUETTE Wing-span 110–132 cm. Size approximately as Buzzard *Buteo buteo*, but wings more equal in width generally, with slightly more ample hand and longer, deeper-cut 'fingers' (five free 'fingers', not four as in *Buteo*). Wings fairly parallel but smooth curve to trailing-edge. Head and neck similar to Buzzard. Rather square-cut tail is long, approximately equal to width of wings. When soaring wings are pressed slightly forward and slightly bowed; when gliding they are lowered a little below level and the hand lowered more; distant silhouette recalling Black Kite *Milvus migrans*.

FLIGHT In soaring, wings are held slightly forward and spread tail is often flexed from side to side. In gliding, wings are angled forward and slightly lowered. Active flight is swift and agile with four to five loose but powerful flaps followed by a glide; wing beats are deeper and stronger than Buzzard's. Stoops often from great heights with closed wings and forward stretched legs.

IDENTIFICATION Of the two distinct colour phases, the light outnumbers the dark by about 7:3 in Europe. Light phase may be confused with pale variants of Buzzard and young pale Honey Buzzards, but all-blackish flight-feathers below with just inner primaries a little paler distinguish it. Young Honey Buzzards in particular can show dark secondaries below but bases are paler and barred, and primaries much paler and dark-barred—never so in Booted. Booted also lacks dark patch at carpal joint, which, when present in light Buzzards, is 'comma'-shaped or larger and in young Honey Buzzard covers most of primary coverts. Tail pattern, kite-like upperwing-coverts, white neck patches are further distinguishing features. With experience Booted can be separated from buzzards by silhouette and flight; the combination of Booted's long, square-cut and sharp-cornered tail, with its low wing-position, elastic wing-beats and ample, long-fingered hand distinguish it. (In Buzzard, tail shorter and more rounded, wings horizontal or raised when soaring and wing-beats stiffer. In Honey, tail-corners rounded, end of tail often slightly notched and sides of tail slightly convex; neck is thinner and wing-beats are wavy.) Adult Egyptian Vulture *Neophron percnopterus* shows same white body and wing-coverts below, but is noticeably larger with more pointed wings and white, wedge-shaped tail.

Dark phase can be confused with Marsh Harrier *Circus aeruginosus* and Black Kite *Milvus migrans*, and perhaps with some dark Buzzards and young dark Honey Buzzards, but is distinguished, apart from silhouette, on pattern of flight- and tail-feathers (see above). Kite-like upperwing pattern and different flight distinguish it from all-dark young Marsh Harriers, which soar and glide on wings lifted in a shallow V. Dark Booted can be difficult to distinguish from Black Kite at distance, especially if tail shape cannot be seen. Soaring Booted has rounded tail, almost square-cut in Black Kite; some Black Kites show rather pale, dark-barred primaries below, in Booted these are blackish with pale inner primaries. Black Kite has much more barred tail and lacks Booted's well-defined, whitish uppertail-coverts and has no white neck patches—all useful points if tail shape is not seen, is damaged or in moult.

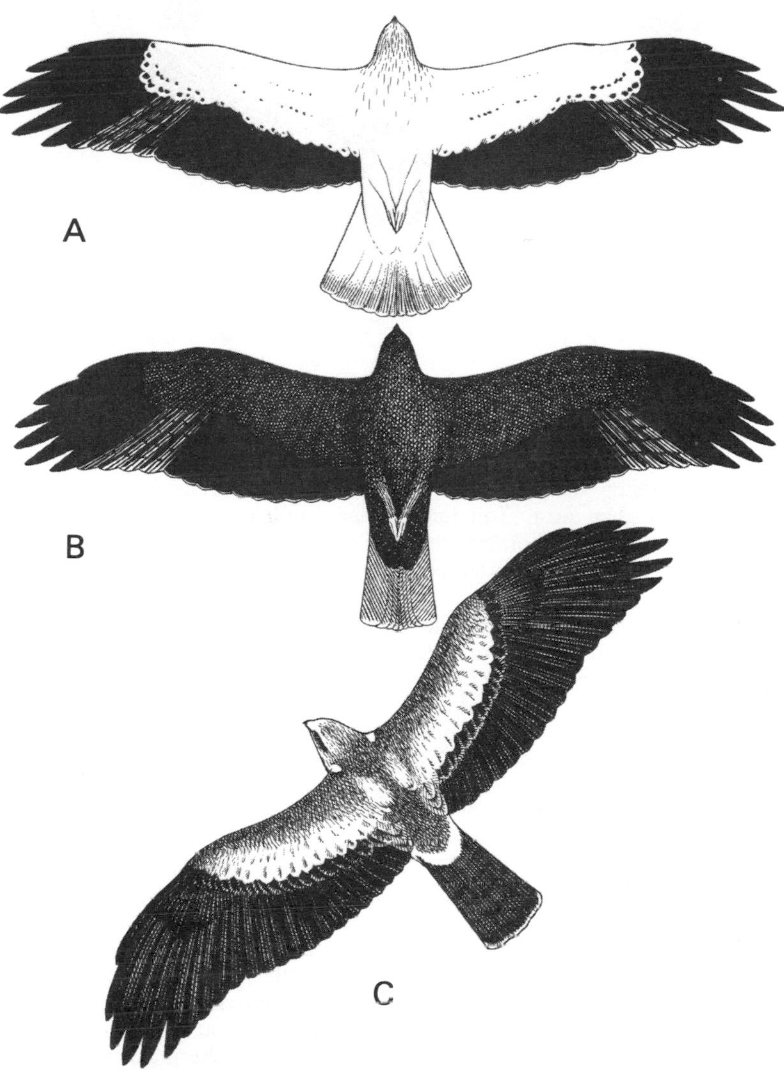

Fig. 26. **Booted Eagle** *Hieraaetus pennatus*. Light phase and dark phase from below, either phase from above. Light phase adult from below (A) has white or creamy-white underwing-coverts with scattered small black spots, particularly on greater coverts. Body white or creamy-white, sometimes washed rusty and black-streaked on lower neck and upperbreast. Blackish-brown patch below eye borders pale throat. Undertail pale brownish-grey with diffuse darker terminal region and creamy tip. Very dark, neat black flight-feathers and pale inner primaries forming more or less conspicuous wedge; translucent white line on trailing-edge of wings can stand out in certain lights. Light phase juvenile very like adult but generally washed more rusty-brown below. Dark phase adult from below (B) has same flight-feathers as light phase and uniform warm brown underwing-coverts (with blackish greater-coverts in some) and body, blackish-looking at distance; brownish-grey undertail paler than body, often with orange tinge when spread and seen in strong sunlight, and narrow dark bar near tip as in light phase. From above both phases and all ages similar, though pale markings paler in light phase generally; conspicuous light yellowish- or buff-brown band through wing-coverts is diagnostic. Scapulars yellow-brown forming light patch on either side of dark brown back. Uppertail-coverts pale brownish- or whitish-yellow, contrasting with blackish-brown rump and uppertail. Blackish-brown flight-feathers show slightly diffuse paler patch on primaries. Where leading upperwing meets side of neck a small pure white spot is visible when bird approaches observer.

Bonelli's Eagle
Hieraaetus fasciatus
(plates 37, 38)

SILHOUETTE Wing-span 155–160 cm. Much larger than Buzzard *Buteo buteo* and approaching Short-toed Eagle in size (compare Figs. 16B and 17A). Head small but noticeably protruding. Wings relatively broad with nearly parallel edges, though hand slightly narrower than arm. Tail longer than breadth of wing and can appear conspicuously long. In head-on profile when soaring, wings more or less flat; when gliding, either flat or slightly bowed like Booted Eagle with hand somewhat lowered.

FLIGHT In soaring, which it does far less readily than most other large raptors, wings are held flat at right angles to body and tail is often closed or only partly spread. In gliding, wings are almost flat with the carpal joints pressed forward in such a way that the leading edges are angled while the rear edges are still nearly straight (carpal joints even more accentuated in stoop, or fast gliding); appearance is then not unlike large Honey Buzzard *Pernis apivorus*. Active flight is swift and agile with powerful, but not deep, wing beats frequently interspersed with glides and occasionally with stoops. Often seen in pairs.

IDENTIFICATION Adult with white underbody and forewings may be confused with smaller light phase Booted Eagle on poor view, but flight-feathers below are clearly paler; blackish tail-band also diagnostic as is white patch on back which can be visible at 1–2 km. Bonelli's does not show such pale and contrasting coverts-band above and lacks whitish uppertail-coverts and white neck-patches of Booted. Shape and length of tail, noticeably more protruding head and shape of wings when gliding also differ from Booted's. Flight silhouette recalls Honey Buzzard's (while Booted's does not), but noticeably larger Bonelli's has thicker head, more sharp-cornered and never notched end to tail, and generally more parallel-edged wings and more deeply cut 'fingers'. Numbers of bars and spacing of these on flight- and tail-feathers also differ (see Figs. 2 and 16). Juvenile Bonelli's without back-patch and conspicuous blackish tail-band is more likely to be mistaken but silhouette is still very useful guide. On plumage, with barred, pale flight- and tail-feathers below, could be taken for a *Buteo*, but flight-feathers below do not darken towards trailing-edge and body and under-wing-coverts are a much more uniform and unpatterned rusty colour. These characters, however, do not hold in all the less well-known varieties of transitional plumage. Again, silhouette is useful guide and, compared to *Buteo* group, the long, narrow tail, rather protruding and relatively narrow neck, longer and characteristically angled wings with more ample hand and more deeply cut 'fingers'.

(*opposite*)

Fig. 27. **Bonelli's Eagle** *Hieraaetus fasciatus*. Three adults (A, B and D) and one juvenile (C) from above. Adult has dark brown greater coverts, scapulars and flight-feathers with slightly paler diffuse patch on primaries; forewing-coverts are dark to medium brown contrasting only slightly with flight-feathers in fresh plumage (A) but can bleach to pale greyish-brown (B). Back dark

brown, with whitish patch which, in most adults, is confined to mantle (A) but can be extensive (B) or merely consisting of a few white flecks (D) which may be young adults. Uppertail grey-brown with five to six very faint (hardly visible in field) narrow bars and broad blackish terminal band. Juvenile (C) resembles adult above but greyish primary patch is usually more conspicuous and barred, darkish brown upperparts lack white patch and grey-brown uppertail lacks dark terminal band, but instead has about eight narrow dark bars, best seen in spread tail. White flecks start to appear on back well into transitional plumage at 1½–2 years of age.

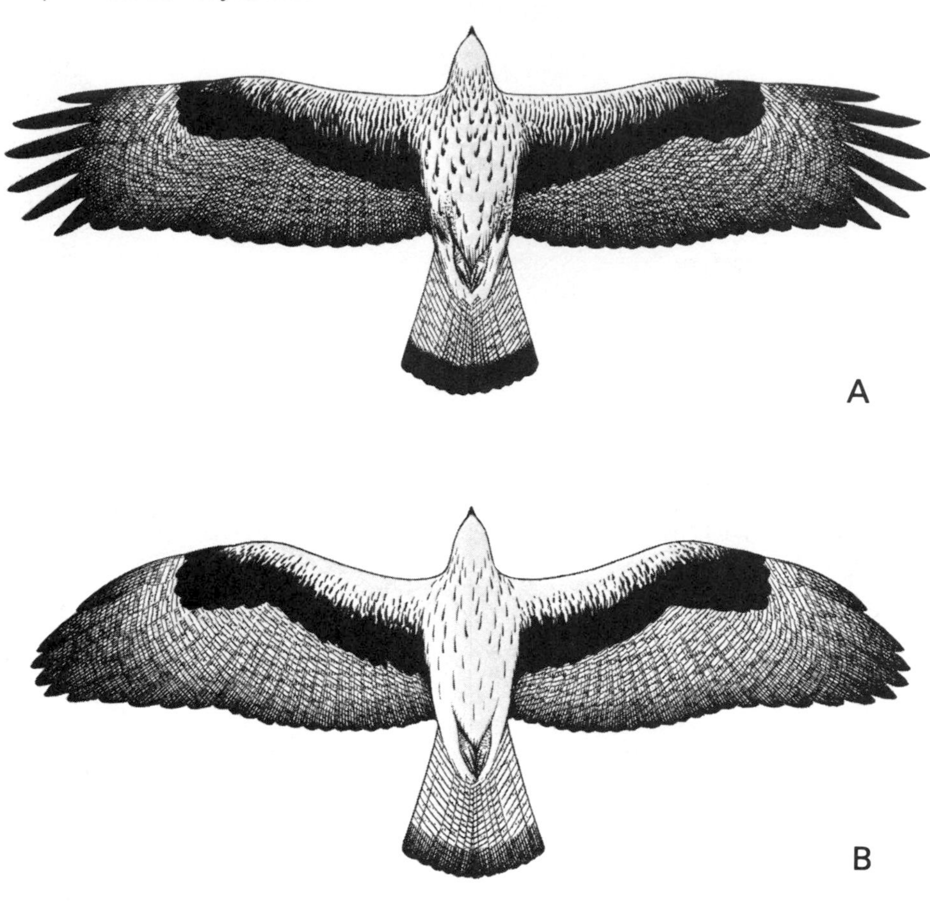

Fig. 28. **Bonelli's Eagle** *Hieraaetus fasciatus*. Two adults (A and B), one immature (C) and two juveniles (D and E) from below. Adult has brownish-white to white underparts, pure white leading-edge to underwings, and body variably black-streaked but usually appearing gleaming white in sunlight. Very dark grey greater coverts or blackish median coverts form diagnostic broad blackish band across underwings, which in some (A) may extend onto lesser coverts (see also Fig. 16B which depicts a bird with virtually all black coverts). Flight-feathers are grey with slightly darker barring and dark trailing-edge which in some birds can form a clear-cut band. A paler area on primaries can be quite conspicuous, but on some individuals (as in B) all flight-feathers appear paler; this feature may be due to sunlight reflecting upwards from pale soil or rocks; wing-tip blackish. At distance, underwings look dark in contrast to whitish body. Pale greyish undertail has clear-cut band at tip. Juvenile (D and E) is much paler bird with body and forewing coverts pale rusty-brown or buff, darkest on upper-breast and merging to rusty-white on undertail-coverts. Tips of greater coverts variable; they can be dark greyish forming diagnostic narrow band through underwing (D) or rusty-brown or buff or even paler (E). Flight-feathers paler than adults' with pale grey to whitish primaries with contrasting dark wing-tip; secondaries slightly darker grey and all flight-feathers are narrowly barred darker (variable in distinctiveness) and without dark trailing-edge of adults. Pale grey undertail is finely dark-barred and lacks blackish terminal band of adult. As a rule, juvenile Bonelli's looks very pale below with or without the dark bar through wing. Immature (C) is highly variable. Body and underwing-coverts may be pale rusty-brown or paler creamy with variable amount of adult-like dark brown streaks or spots and apparently all birds in transitional plumage show dark grey coverts band. Example (C) shows mixture of juvenile and adult flight- and tail-feathers which will shortly give way to all-adult feathers.

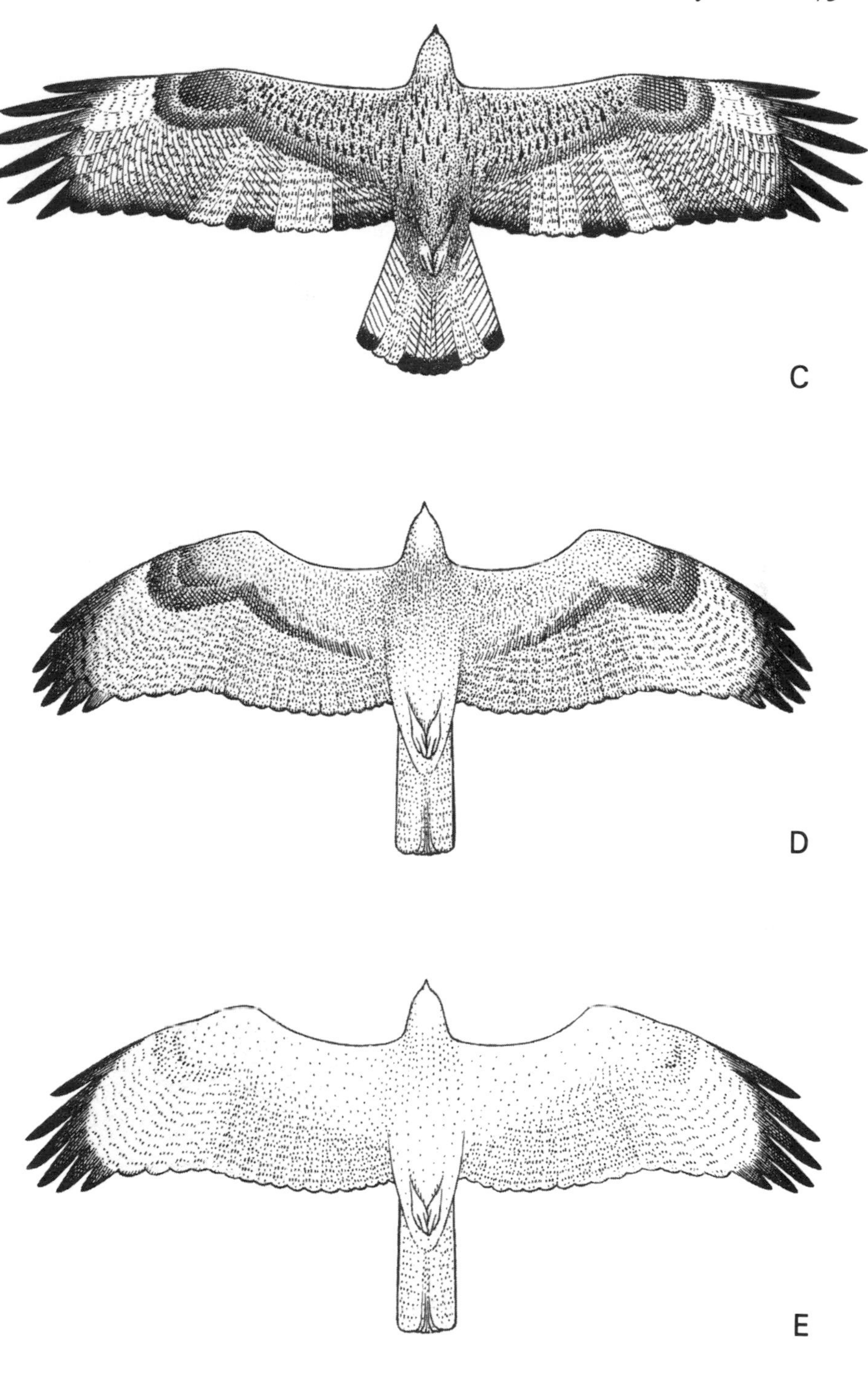

Fig. 28. (cont.). Bonelli's Eagles from below.

Short-toed Eagle
Circaetus gallicus
(plates 39–41)

SILHOUETTE Wing-span 185–195 cm. Medium-large eagle, a little larger in wing-span than Osprey or Lesser Spotted Eagle. Head broad and well protruding. Wings long, especially arm, and moderately broad, smoothly S-curved at trailing edge from relatively narrow wing base to greatest width of wings at carpal joint level, a feature that gives wings at times a resemblance to Lapwing *Vanellus vanellus*. When closed, medium long tail is relatively narrow with sharp corners and square-cut at end. When soaring, wings pressed slightly forward and carried more or less flat but sometimes slightly raised and 'fingers' even more noticable than in *Hieraaetus* and *Buteo*. When gliding, carpal joint is pressed well forward almost to bill level with trailing-edge at right-angles to body, not unlike gliding Honey Buzzard; but when gliding fast both wing edges are angled. Seen head-on when gliding, arm slightly raised and hand correspondingly drooped.

FLIGHT Active flight has slow wing-beats making bird look very big; wing-beats are deep, powerful and elastic, rather like those of Golden Eagle; in interspersed glides tail sometimes twisted in kite-like fashion, though less frequently, and glides are less steady, strong and straight than in *Aquila* or *Hieraaetus*. Except on migration spends much time hovering in search of prey with shallow slow wing-beats and dangling legs.

IDENTIFICATION Likely to be confused only with Osprey, pale Honey Buzzard and pale Buzzard. Osprey has narrower and proportionately longer wings which are more angled and gull-like; it also has dark carpal-patches, dark band through middle of wing and otherwise gleaming white underwing-coverts and underbody, whereas Short-toed's barring, though sometimes very fine, makes it at best very pale rather than gleaming white. At closer ranges, characteristic head shape, head markings and tail pattern of each species can easily be seen. Dark-headed, heavily-barred Short-toed can recall pale adult Honey Buzzard but lacks latter's usually distinct dark carpal patch, black primary 'fingers' and has different spacing of darker tail bands. From Buzzard and young Honey Buzzard on lack of dark carpal-patch and darkish wing-tip ill-defined from whitish primaries below; in majority of Buzzards wing-tip blacker and more sharply demarcated. Short-toed's larger size, longer-armed wings, square-cut and sharp-cornered tail and slow wing beats also distinguish it from Buzzard and Honey Buzzard.

Fig. 29. **Short-toed Eagle** *Circaetus gallicus*. From above. Most plumages show distinct contrast between darkish grey-brown flight-feathers and paler greyish to buff-brown coverts, but there is much variation as indicated by illustrations. In typical birds, pale coverts are darker on leading-edge and on greater coverts (B and C) whilst in others entire coverts are dark grey-brown (A) and a few can be sandy-buff, or even paler, superficially resembling Booted Eagle (D). Thin whitish line sometimes seen on tips of greater coverts and uppertail-coverts (juvenile?) and can be seen on most of types previously described. Colour of mantle and back usually as wing-coverts, though often darker, particularly on lower back to rump; hind-neck to top of head often paler than rest of upperparts. There is a diffuse palish area on primaries above, varying in distinctness, but normally contrasting with darker primary tips. In all types three (occasionally four) dark bands are visible on brownish tail but at distance closed tail usually shows just darker terminal band. Thin whitish line on trailing-edge of wings and tail in fresh plumage.

Fig. 29. (cont.). Short-toed Eagles from above.

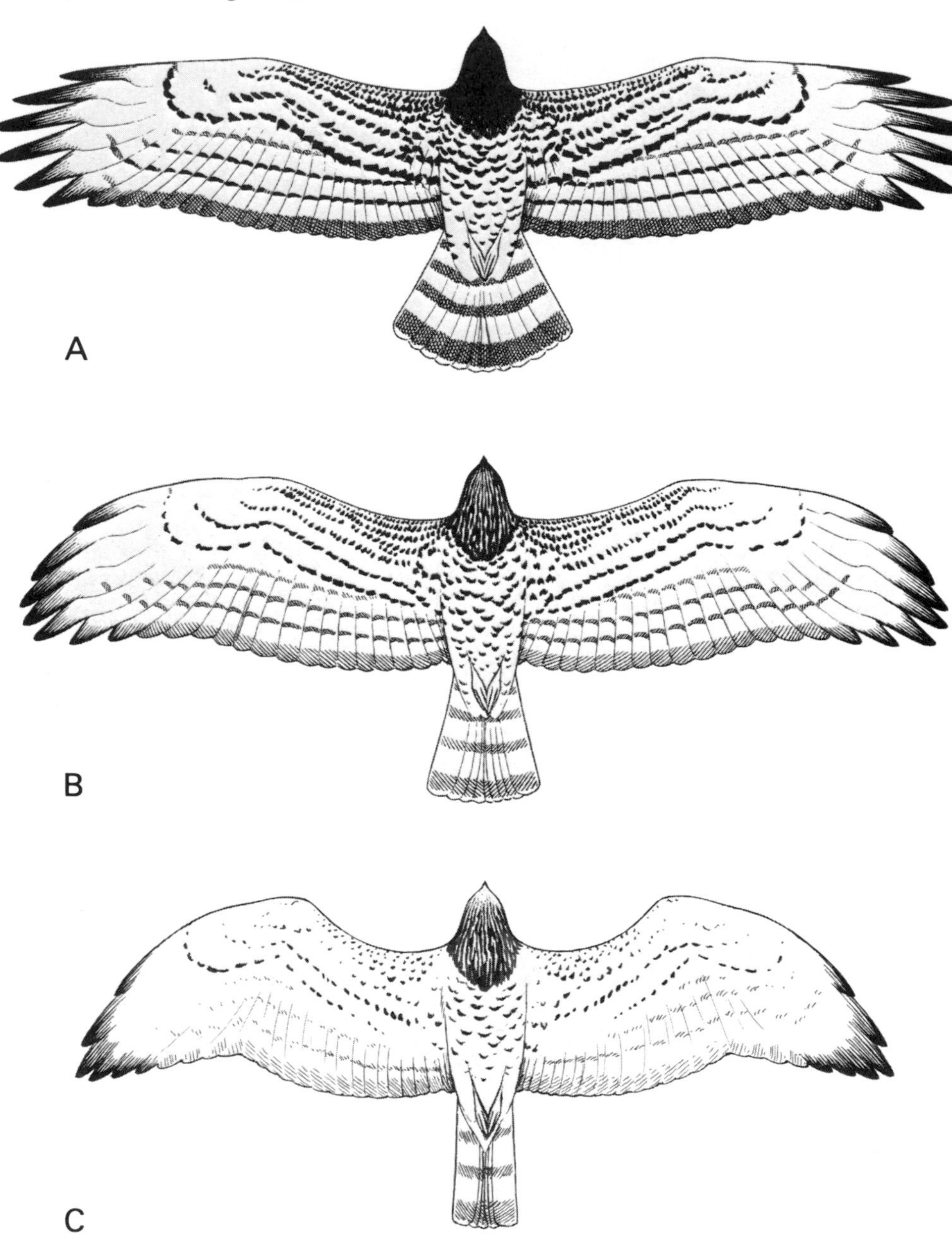

Fig. 30. **Short-toed Eagle** *Circaetus gallicus* from below. The basic colouration is white and at a distance most individuals look off-white underneath with a distinct dark hood and dark-grey tips to the primaries (30B, C). some have very dark chocolate hoods and much stronger blackish-brown markings on the wings and underbody (30A), the latter recalling the pattern of a Honey Buzzard *Pernis apivorus*. Others, but these are the least common, have white heads and almost unmarked wings (30D) apart from the dark-grey tips to the primaries, which remain in all plumages. Three bands on the tail can be seen in even the most lightly marked individuals. The cline of plumages from those with very dark heads and heavy patterning to those with white heads and few markings is not properly understood, though it may be linked with age.

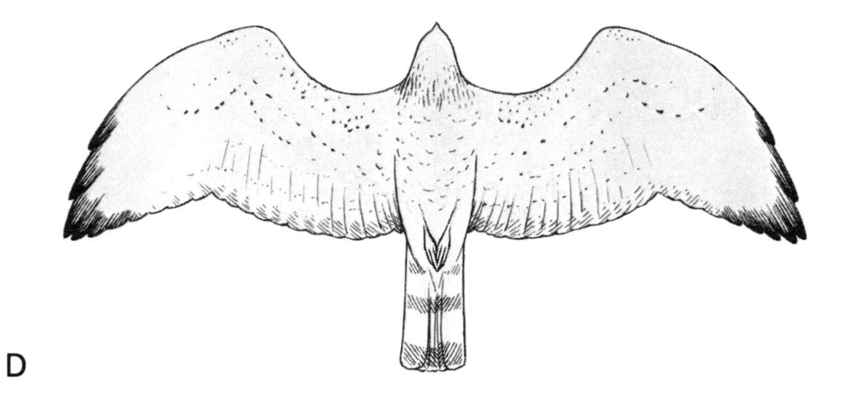

Fig. 30. (cont.). Short-toed Eagle from below.

Osprey
Pandion haliaetus
(plates 42, 43)

First summer Herring Gull with Osprey

SILHOUETTE Wing-span 145–170 cm. Medium-large, long-winged raptor with medium-short tail. Wings in most positions are decidedly angled and hands look long and narrow, giving appearance not unlike large gull *Larus sp*. Head fairly small, but neck well-protruding with upward curve. In head-on profile when gliding, wings sharply bowed in manner of large gull, less so when soaring.

FLIGHT When gliding, arm is raised and hand lowered, thus giving bowed appearance to wings seen from head-on; from below both fore- and trailing-edge of wing angled with carpal joint pressed well forward. Wing-beats rather loose and shallow, but powerful and steady, and interspersed with long periods of gliding. Captures fish by hovering over water with loose heavy wing-beats and dangling legs, or by gliding then diving head first, with feet thrown forward and down, as it submerges with a huge splash.

IDENTIFICATION Confusion hardly possible with any other raptor except Short-toed Eagle, and even then Osprey is noticeably smaller with a narrower head and thinner and proportionately longer wings which are more angled and gull-like; it also has black carpal patches and a dark band along the middle of the wing, while the underbody and underwing-coverts are gleaming white rather than, at best, very pale or whitish. Observers should beware, however, of possibility of overlooking distant Osprey as large gull, the shape and basic colouration being very similar.

Fig. 31. (*Opposite*). **Osprey** *Pandion haliaetus*. Adult and juvenile very similar from below (A): throat white bordered by black where stripe through eye extends on to the sides of the neck; underbody and underwing-coverts white, often strikingly so, except for the broad but variable band of buff and brown streaks (less defined in the juvenile) across the upper breast and the wide band of blackish along the rear edge of the coverts in the middle of the wings; the last joins the black carpal patches and further contrast is provided by the black wing tips; otherwise the primaries, secondaries and tail are narrowly and indistinctly barred with grey. In contrast to the largely white underparts, the upperparts are mainly dark brown. Adult from above (B) is uniformly dark except for the white head and slightly paler tail. Juvenile from above (C) is generally lighter brown and speckled with creamy white. Tails of both adult and juvenile become translucent when spread, particularly seen against the light, and can then seem whitish or slightly rufous.

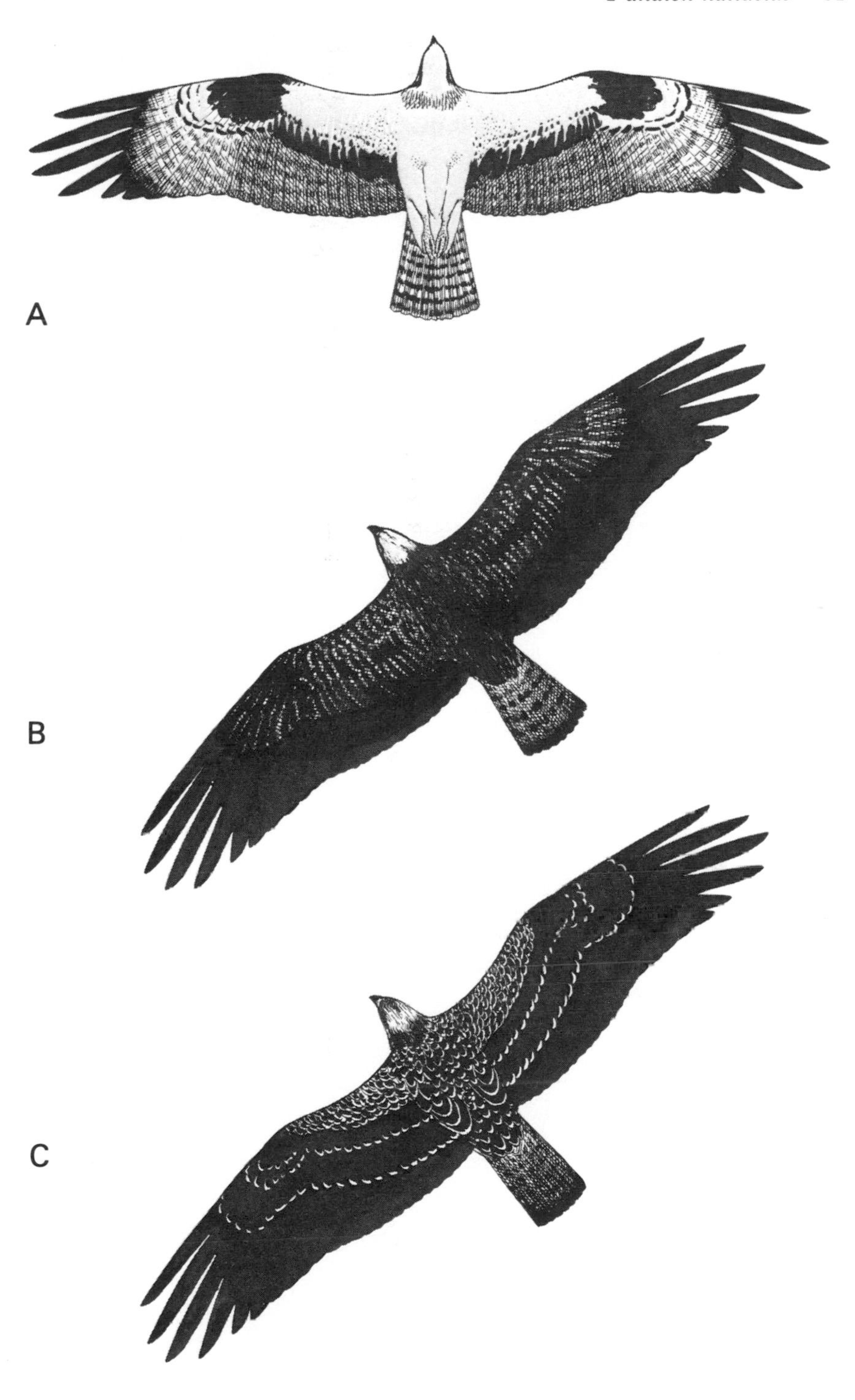

Fig. 31. Osprey from below and adult and juvenile from above.

3 The Harriers and Kites

The four harriers *Circus* form a genus associated with extensive reedbeds. cornfields and moorland, not requiring trees or cliffs as do most raptors. Three—Marsh *C. aeruginosus*, Hen *C. cyaneus* and Montagu's *C. pygargus*—are comparatively common, but the Pallid *C. macrourus* is much rarer (mainly east Europe). All invariably soar and normally glide with wings raised in a shallow V, and this and their long wings and tail distinguish them from other birds of prey. The only other similar-sized raptors to soar on raised wings are the buzzards, particularly *Buteo buteo*, but they have short tails more fully spread and glide on flat wings. Male harriers are not difficult to identify with reasonable views, but the females and immatures of Hen, Montagu's and Pallid must often be grouped as 'ringtails' unless the head and neck pattern is seen.

The two kites, because of their forked tails (much less obvious in Black Kite), are usually not difficult to identify. The Black Kite may resemble a dark phase Booted Eagle or immature, or female, Marsh Harrier but the latter is readily identified by its raised wing glide.

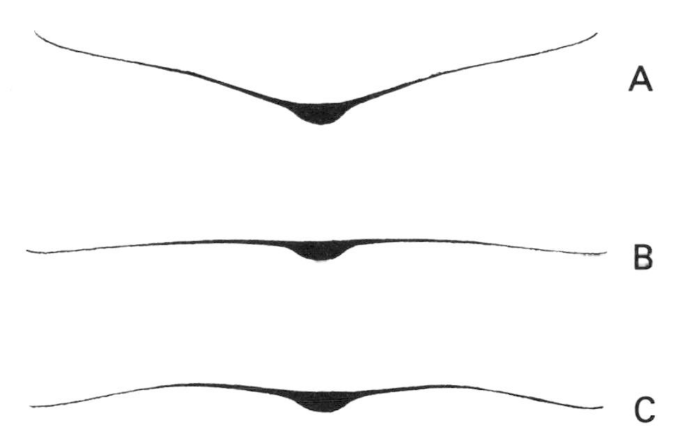

Fig. 32. Head-on profiles of harriers and kites; (A) typical soaring and gliding of harriers with wings in shallow V; all species of harriers, particularly the Hen Harrier *Circus cyaneus*, occasionally hold the wings flat or even slightly bowed (B), but only when gliding, never when soaring. The kites (C) soar and glide on slightly arched wings.

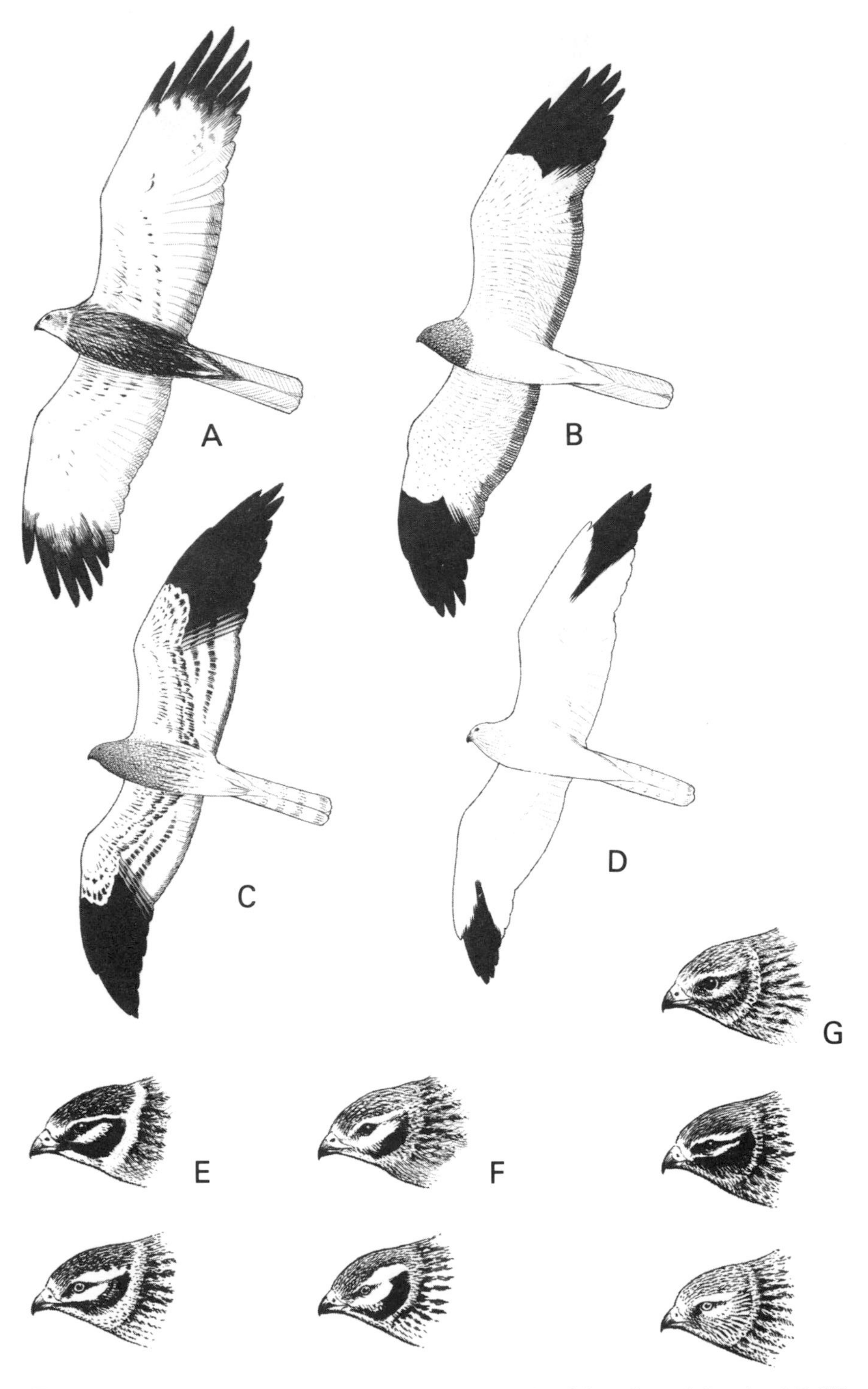

Fig. 33. Typical undersides of male harriers and heads of juvenile and female of Pallid Montagu's and Hen.

Fig. 33A. **Marsh Harrier** *Circus aeruginosus*. Old male with extreme pale underwings shown. Largest harrier, with rather broad body, long and fairly broad wings, and moderately long tail. Male rufous-brown below, with paler head, grey tail and white to pale grey underwings except for black ends to primaries and brownish-buff, occasionally whitish coverts; female and immatures entirely chocolate-brown, but often (especially adult female) with yellowish crown, throat and leading edge to wings. Flies with rather heavy wing beats. Fairly widespread in large reed-marshes in Europe, north-west Africa and Middle East, but absent Iceland, Ireland and all except south Fenno-Scandia (where local) and nearly extinct Britain; those from north-east and central Europe move south September-April and some migrate to tropical Africa.

Fig. 33B. **Hen Harrier** *Circus cyaneus*. Adult male shown. Shape similar to Marsh, but less bulky and with narrower wings. Male easily identified by white underparts with mid-grey wash on head and upper breast, all black primaries and dark trailing edge to underwings, also upperparts clear grey with black primaries and white rump; slightly larger female streaked brown with bars on primaries, secondaries and tail, and white rump (see also 33G). Breeds widely in moorland, large fields and open marshes in north and central Europe south to north Iberia, north Italy and Caucasus, but not Iceland, most of Iberia and Italy nor Balkans, and in Britain largely confined to Scotland, including Orkney and Hebrides; northern populations move south October–April when fairly common in marshland over much of central and south Europe and Turkey.

Fig. 33C. **Montagu's Harrier** *Circus pygargus*. Adult male shown. Smaller than Hen with slimmer body, narrower wings and more buoyant flight; male slightly larger than male Pallid. Male dirtier-looking than Hen or Pallid with dark grey upperparts, head and upper breast, rusty streaks on flanks and underparts, black primaries, ill-defined dark bands on undersecondaries and narrow black bar on upperwing; female similar in plumage to female Hen (but see 33F); both sexes have melanistic form. Summer visitor to open marshland, moors, heaths and cornfields in Europe and north-west Morocco, but absent Iceland, much of Fenno-Scandia (except south Sweden and Denmark), most of Italy and Balkans, irregular Scotland and Ireland, and now very few England and Wales; migrants most regular south Europe, north Africa and Middle East in May and late August-early September.

Fig. 33D. **Pallid Harrier** *Circus macrourus*. Adult male shown. Male slightly smaller than male Montagu's with slimmer body, narrower wings and lighter, more agile, almost tern-like flight which, together with pale grey upperparts, white underparts and narrow wedges of black at wing-tips, recalls Common Gull *Larus canus*; female similar in size, shape and plumage to female Montagu's and virtually impossible to distinguish unless whitish collar visible (see 33E). Summer visitor to steppes and plains in central and south Russia south to Caucasus, west to Romanian Dobruja, and has extended erratically to Sweden (Gotland and Öland) and Germany; migrants most regular east Balkans, Turkey and Middle East in April and September–early October (thus earlier in spring and later in autumn than Montagu's), while a few winter south-east Europe and Italy.

Fig. 33. E, F and G. Typical head patterns of juvenile and female 'ring tail' harriers. E. **Pallid.** Upper juvenile; lower adult female; note whitish collar behind dark crescent on cheeks and well-defined eye-stripe. F. **Montagu's.** Upper juvenile; lower adult female; note similar pattern to Pallid but almost no pale collar and only faint black line through eye, thus side of face looks whiter. G. **Hen.** Upper juvenile; centre juvenile (dark-faced type); lower adult female; note usually streaked and rather owl-like face pattern with faint narrow collar bordered by ruff of streaks extending to breast.

Fig. 34A. **Red Kite** *Milvus milvus*. Adult shown. Rather slim, rakish build with long, angled wings and long, well-forked tail. Pure white wing-patches usually appear in startling contrast to blacks, browns and rufous-browns; head also normally much whiter than Black Kite, but immature can be confused with well-marked immature of that species. Flight very buoyant with tail constantly manoeuvred; and soars on slightly arched wings. Central Wales, most of continental Europe east to western Russia (but only south Sweden in Fenno–Scandia and not north France to Holland or south Balkans), Caucasus to north Iran, and north-west Africa; only March–October in northern parts of range except for sedentary population in Wales and to some extent Sweden, remainder moving south to winter in Mediterranean region including north Africa, Middle East and south Balkans.

Fig. 34B. **Black Kite** *Milvus migrans*. Adult shown. Basically much dingier and more compactly built than Red Kite with tail only slightly forked. Pale primary patches variable, but typically as shown; immature can be confused with Red Kite. Flight similar to Red Kite though lacking same grace and finesse; often patrols lakes and rivers, scooping offal and dead fish from water with feet. Most of Europe, April–September, except Britain, Ireland, north France, Fenno-Scandia and south Greece; also North Africa, Turkey and Middle East; passage through whole Mediterranean area, mainly in April and August–September; most winter in Africa, but a few as far north as Balkans and west Turkey.

Marsh Harrier
Circus aeruginosus
(plates 44, 45)

SILHOUETTE Wing-span 115–130 cm. Largest harrier, about size of Buzzard *Buteo buteo* but with slimmer head and body. Wings relatively long and parallel-edged with fairly rounded hand for a harrier. Long narrow tail (about width of wings) and with slightly rounded end. When soaring and gliding, wings raised in shallow V, frequently with band at carpal joint.

FLIGHT When soaring and gliding, wings are invariably raised. When seen gliding from below wings clearly angled with carpal joint pressed forward to head-level and closed long tail looking narrow. Active flight a series of 5–10 rather heavy wing-beats, followed by glide on raised wings, sometimes with long legs dangling. Except on migration, flight is typically low over ground or reed bed, though in spring performs aerial displays high over breeding area.

IDENTIFICATION Can be distinguished from other harriers by larger size, broader wings with more rounded tips and comparatively shorter tail, though silhouette not strikingly different from Hen Harrier. Old male with entirely pale underwings may superficially resemble males of other species, but Marsh always has dark body below and dark back and wing-coverts, whereas these areas are basically white and pale grey in other three. Female or immature with varying amounts of pale (usually yellowish) on top of head, throat and forewing are also easily identified, but all dark brown individuals which, however, have paler bases to primaries below, and slightly gingery tails, are more difficult to identify; when one of these is seen at distance or high overhead on migration, confusion can arise with dark Buzzard, dark Honey Buzzard *Pernis apivorus*, Black Kite *Milvus migrans* and dark phase Booted Eagle *Hieraaetus pennatus*. Last three species, however, always glide and soar on flat or flattish wings, and Marsh can thus be readily distinguished. In addition, Honey Buzzard has narrower and more protruding neck and rounded tail corners; Black Kite has forked tail and Booted Eagle (which see) has more sharp-cornered and quite square-cut tail. Buzzard and Honey Buzzard also have much paler primaries below. Best told from distant Buzzard, which also soars on raised wings, by little longer and less fanned tail and slightly narrower, more parallel-edged wings.

Fig. 35. **Marsh Harrier** *Circus aeruginosus*. Adult male (A), Juvenile (B) and adult female (C), from below. The adult male has a pale yellowish-white or greyish-white dark-streaked head, a buffish underbody more or less heavily streaked with dark russet (looking evenly dark at a distance), white flight-feathers, usually, but not always with a dark trailing edge to wings (most evident on secondaries) well-marked blackish tips to primaries and a very pale grey or even whitish tail. Underwing-coverts are variable, either russet-brown white without any dark markings. A few adult males are completely white to underwings, tail and also head and breast, the only dark colouration being on lower belly and vent and wing-tips, thus very much resembling male Hen Harrier. The adult female is chocolate-brown (paler, more greyish in worn plumage) often appearing darker on the wing-coverts, with paler flight-feathers or just paler primaries, a brown tail often with a warm russet cast. The throat, crown and nape are usually yellowish or creamy-white, but few do not have any pale pattern on the head, thus resembling some juveniles. The chocolate-brown underbody has a variable patch of yellowish on the breast, varying in size and distinctness. Some adult females (worn plumage?) do resemble males a little, since flight-feathers may become greyish-brown, but difficulties should not arise. Few adult females show some yellowish to leading edge of the under-arm. The juvenile from below averages clearly darker, almost blackish-brown in the field, being somewhat paler on flight-feathers below, especially primary-bases. It normally lacks the yellow patch to breast and forewing, seen in adult female; light head markings often (but not always) reduced to pale nape patch or just pale ochre-yellow feathering to forehead and throat and birds with all dark heads are not uncommon. Second-year birds resemble juveniles, but may show some yellowish to breast and leading edge of forewing. Sub-adult males have flight-feathers more or less greyish and the dark wing-tip shows up well; some still have yellowish throat, nape from immature plumage. (See also photographs.)

Fig. 36. **Marsh Harrier** *Circus aeruginosus*. Adult male (A), adult female (B) and juvenile (C) from above. The adult male is tricoloured with dark brown back and wing-coverts (apart from primary-coverts, outermost greater-coverts and leading edge to arm), black wing-tips and otherwise pale grey to greyish-white primaries (sometimes with darker greyish hue), secondaries, primary-coverts, outermost greater-coverts and tail (the last with obscure bars on the outer feathers). In most fully adult males uppertail-coverts are more or less white, sometimes producing a white rump patch. The upperhead is pale, sometimes whitish streaked darker, and a variable, sometimes very large, creamy-yellow patch on forewing. The adult female is variable, medium brown, grey-brown or chocolate-brown to wing-coverts, with the shoulders and lesser wing-coverts yellowish or creamy-white forming variable patch to leading edge of wings; throat and crown similarly coloured (separated by dark line through the eye). Flight-feathers are dark brown or greyish-brown with slightly darker wing-tips and sometimes diffuse darker trailing edge to wings. Tail is slightly paler, tinted rusty. The juveniles average darker brown with often more contrasting head markings (see photograph) and pale tips to greater-coverts but occasionally without any ochre-yellow at all. Second year birds resemble juveniles, but may show yellowish patch on forewing. Sub-adult males have flight-feathers more or less greyish, dark wing-tips, dirty grey primary-coverts and tail; in younger individuals, yellowish forewing, throat and nape and a dark subterminal band to tail may be seen.

Hen Harrier
Circus cyaneus
(plates 46–48)

SILHOUETTE Wing-span 99–121 cm. Smaller, slimmer and narrower-winged than Marsh, but has same rounded ample wing-tips; larger- and broader-winged than Montagu's and Pallid, and thus appearing shorter-winged than either species. Narrow tail clearly longer than width of wings, slightly rounded at end. When soaring and usually when gliding, wings raised in shallow V, often with slight bend at carpal joint.

FLIGHT When soaring, wings raised. When gliding, wings usually raised, but

Fig. 37. **Hen Harrier** *Circus cyaneus*. Adult male, second-year male and female from below. Adult male (A) is white with well-defined grey throat and upperbreast, black 5–6 outermost primaries, a dark grey trailing edge to the wings (conspicuous at moderate ranges, but less noticeable in older birds) and a pale off-white and unbarred tail; when seen overhead in strong sunlight, translucent white secondaries can give appearance of tricoloured wing (plate 47). Adult female (B) very similar to Montagu's and Pallid females, with ground colour of buff-brown, finely streaked on underparts and under-wing with blackish-brown; broad brown barring on primaries and secondaries and about five bands on tail; head patterns lack more contrasting patterns of Montagu's and particularly Pallid (see Fig. 33 especially). Juvenile (not shown) is very similar to adult female, though small differences (variable in distinctiveness) are: generally streaked underparts having more rufous ground-colour, secondaries averaging darker, below, as in other juvenile 'ring-tails', and dark crescent on ear-coverts usually darker. Occasional young birds, at least until first winter, have surprisingly bright rufous ground-colour and dark secondaries below. Sub-adult male (2–3 years old) differs from adult in having long, narrow, rufous-brown, scattered streaks in white underparts below grey upperbreast, and in some, fine dark bars on secondaries below and grey speckling on greater underwing-coverts. Example of male in early transition (C) depicts second-year bird (1+ years old) which shows new inner seven primaries (adult male-like), whilst remainder of flight-feathers are juvenile: pattern being not unlike that of male Pallid. Note also moulting tail and white underparts with juvenile streaks on upperbreast; few grey feathers are seen on neck, face and upperbreast at this stage.

occasionally flat. Active flight similar to Marsh, but wing-beats faster and interspersed glides generally shorter.

IDENTIFICATION Male readily distinguished by white underparts with grey throat and upperbreast, extensive black on primaries, dark trailing-edge to wings, wholly pale grey upperparts with solid black primaries and white uppertail-coverts. Male Montagu's is rather darker-looking on secondary coverts above in contrast with silvery grey primary coverts; one black band along coverts above and two below. Male Pallid is much whiter on head, throat and upperbreast, lacks well-defined white patch on uppertail-coverts (though can be seen in sub-adults), has much less extensive black on wing-tips and lacks blackish band on rear edge of underwing. Broader, comparatively shorter wings of Hen with less pointed tips (formed by 2nd-5th primaries; in Montagu's and Pallid by 2nd–4th) and less buoyant, less tern-like flight distinguish female with practice from female of other two (but beware of moult in outer primaries). Female Hen has on average more white on uppertail-coverts than females of other two, but this is not a useful field character. At close range, head pattern may aid identification: female Hen has less distinct dark patch on ear-coverts than females of other two and has narrow pale collar, which is usually lacking in Montagu's, and is much broader in female Pallid. Juvenile Hen (greatly resembling female) is distinctly streaked from owl-like ruff, thickly across breast and sides of neck and more thinly onto underparts. Only juvenile European harrier to be streaked on underbody, and usually less contrasting head markings also aid identification (see Fig. 33); however, there is some recent evidence to suggest juveniles may, on occasions, have unstreaked rufous underparts as in Montagu's and Pallid.

Fig. 38. **Hen Harrier** *Circus cyaneus*. Adult male and female from above. Adult male (A) is grey, even slightly blue-grey, palest on flight-feathers, slightly darker on coverts and darkest on back. Dark trailing-edge to wing on younger birds is gradually lost with age. Outer six primaries black, and clear white patch on uppertail-coverts. Adult female (B) has dark brown back and wings with paler area on primaries traversed by broken dark barring. Buff tinge to median coverts forming pale area towards fore-edge of wing, usually only noticeable in good light or at closer ranges. White uppertail-coverts (typically larger than those of female Montagu's and Pallid) and grey to grey-brown tail with broad bands of dark brown. Head pattern normally lacks contrast of female Montagu's and Pallid, but this feature useful only at close range (see also Figs. 33 and 37B). Juveniles virtually indistinguishable from adult female, but very often noticeably brighter, deeper rufous on body and wing-coverts and darker on secondaries (see also caption to Fig. 37).

Typical habitats of the four harriers. It must be emphasised that much overlap occurs, especially in winter and on migration.

Montagu's Harrier
Circus pygargus
(plates 49–51)

SILHOUETTE Wing-span 97–115 cm. Male noticeably slimmer and narrower-winged than male Hen (compare Figs. 33B and 33C): as a result wings appear longer and, at a distance or high overhead, shape and structure not unlike long-tailed falcon, such as Kestrel *Falco tinnunculus*. Size corresponds to Pallid, although males of that species are often slightly smaller. Female appears a little larger than male, though size difference is, in fact, very small. In head-on profile when soaring or gliding, wings raised in shallow V.

FLIGHT When soaring or gliding, wings raised, often with bend at carpal joint and in higher V than Hen. Active flight is very light and bouyant, usually 5–6 leisurely wing-beats followed by a wavering glide; tern-like, rhythmically rising and falling of body very characteristic, particularly in lighter male. Generally low over ground, but often soars high in breeding season, when V of wings becomes accentuated. Like other harriers, will suddenly check in flight and turn agilely to pounce down to ground after prey.

IDENTIFICATION Male smaller than Hen, being slightly larger than Pallid. Readily distinguished from either by darker, less precisely patterned appearance with streaks on flanks, black secondary-bands on underwing and darker grey upperparts with narrow black bar on upperwing, all features lacking in paler and cleaner-looking Hen and Pallid. Female smaller with narrower, comparatively longer wings and more pointed wing-tips than in female Hen (which see) and very similar in size, proportions and colouration to female Pallid. However, more widely spaced blackish secondary-bands below are just visible above, whereas female Pallid's darker bands are not. Lack of conspicuous pale collar in female Montagu's aids identification (see Fig. 33). Juvenile's lighter build, proportions and flight, unstreaked rufous underparts and underwing-coverts distinguish it from brown Hen Harrier. May be told from juvenile Pallid by head pattern: no distinct pale collar behind dark crescent on ear-coverts which do not reach base of bill; less distinct or no black eye-streak, more white above eyes generally, but not so dark on sides of neck as is usual in juvenile Pallid. Dark colouration on secondaries below (a non-adult character) and plumage of underparts and underwing-coverts in juvenile plumage (from deep chestnut to yellowish-orange or even creamy) overlap completely in the two species, though juvenile Pallid is usually palest below. Second-year males in spring (about ¾ year old) still with juvenile flight- and tail-feathers can be difficult since Montagu's may show pale collar at this stage; usually, however, it has more or less greyish cheeks and darkish grey throat and upperbreast in contrast to very pale remainder of underparts (second-year spring male

(*continued p. 96*)

Fig. 39. **Montagu's Harrier** *Circus pygargus*. Adult male, adult female, transitional male, melanistic female and juvenile from below. Adult male (A) has dark ash-grey head to throat and breast, smudging to rufous-streaked whitish belly and flanks. Outer six primaries black, four inner greyish becoming whiter inwards; secondaries white with two black bands and narrow dark grey trailing-edge. White wing-coverts flecked black on primary coverts, streaked rufous on greater coverts and barred rufous on axillaries and some fine streaks on lesser coverts. Pale grey undertail is barred darker. Adult female (B) resembles female Hen Harrier but spacing between black secondary bands wider and dark crescent on ear-coverts generally more distinct, but no pale collar. Juvenile (E) resembles female but has an unstreaked, warm chestnut (most usual colour) to yellowish-orange underbody and wing-coverts (dark streaking on primary and greater coverts). Colour of secondaries varies from dark slate-grey (with darker bands almost invisible) to much paler whitish with dark bands; note: dark secondaries below suggest a young bird. Both sexes have a rare melanistic form. Male (not shown) is sooty-blackish below with or without silvery area at base of primaries and tail; female (D) is dark chocolate-brown except for greyish-white bases to primaries, which are either unbarred (as shown here), or marbled or strongly barred darker; may give appearance of small, slim-winged, light-flying female Marsh Harrier. Young males develop adult male-like flight- and tail-feathers during their second autumn (i.e. 1+ years old); the one illustrated (C) is at a slightly younger stage, having a mixture of adult male and juvenile flight-feathers—giving male Pallid Harrier-like pattern. Darkish grey head and upperbreast (but often with pale collar!) characteristic of pre-one-year-old birds and white rufous-streaked underparts and underwing-coverts characteristic of birds after this period.

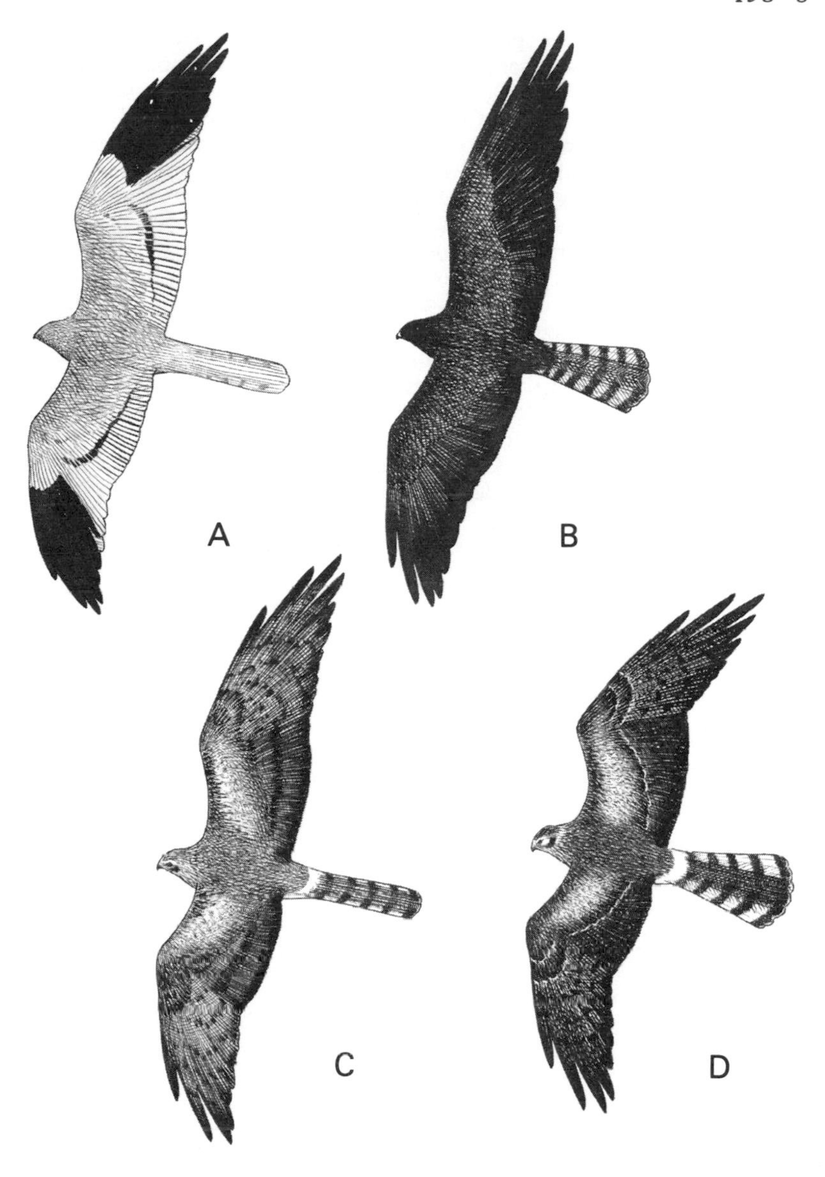

Fig. 40. **Montagu's Harrier** *Circus pygargus*. Adult male, melanistic female, adult female and juvenile from above. Adult male (A) is dark bluish-grey on head, upperbreast and most secondary coverts, contrasting with much paler silvery-grey primary coverts, and sometimes inner primaries with which black outer primaries merge. Black band at base of secondaries and sometimes inner primaries. Closed tail unbarred ash-grey (central pair), remaining feathers distinctly banded. Uppertail-coverts often almost entirely grey, only seldom with narrow white patch. Melanistic female (B) is entirely dark chocolate (charcoal grey in male), apart from banded tail which male lacks. Adult female (C) is darkish grey-brown with creamy-buff median and some lesser coverts, paler primaries (except for dark tips and barring) than secondaries, white patch on uppertail-coverts and banded tail. Blackish secondary bands below are just visible above at close range. Head pattern includes a dark crescent on ear-coverts but only a faint or no black line behind eye and no whitish collar (Fig. 33). Juvenile (D) resembles female above but has darker secondaries and upperparts, generally with narrow pale line along tips of greater coverts, has more rusty tinge to coverts and frequently a pale nape spot.

Pallid has whiter cheeks, more pronounced dark eye-streak and narrow pale collar—juvenile-like pattern—and normally much paler throat and upperbreast with less or no contrast to pale underparts); spring male without white uppertail-coverts (moulted early in spring generally), but nearly all-grey tail-coverts, is a Montagu's. Second-year summer/autumn male (1+ years old) with flight- and tail-feathers in moult, distinguished from similar-aged Pallid by blacker renewed primaries above, two partial black bands on secondaries/inner primaries below and sometimes visible above. Montagu's also shows distinct rufous streaks on lower breast-belly and barred axillaries and rufous-streaked greater underwing-coverts, while Pallid is much softer and less conspicuously streaked rufous-brown on under-parts and lacks marks on axillaries and greater coverts; dark grey upperbreast, in contrast with whitish remainder of under-parts in Montagu's, and grey sides to head (but perhaps pale collar) in contrast with whiter cheeks and upperbreast in Pallid, is an even more pronounced difference at this stage. All, or nearly all, grey uppertail-coverts only seen in Montagu's. When second plumage is nearly fully assumed (almost two years old), male greatly resembles adult male but black secondary-band above can be very diffuse or almost lacking (but present below). Rare melanistic form easily identifiable, but needs to be distinguished from all-dark female or immature Marsh, which is, however, larger and broader-winged.

Pallid Harrier
Circus macrourus
(plates 52, 53)

SILHOUETTE Wing-span 99–117 cm. Very similar in size, proportions and general outline to Montagu's (which see) but with clearly narrower and more pointed wings and slimmer body than Hen. Male slightly smaller than male Montagu's but females of the two are of similar size. In head-on profile when soaring or gliding wings raised in shallow V, on average less high than in Montagu's; also glides, often with lifted arm and level hand or even on nearly flat wings.

FLIGHT Active flight like Montagu's: light, buoyant and tern-like, particularly the smaller male.

IDENTIFICATION Safe distinction between Pallid and Montagu's on flight silhouette probably not possible, though slightly shorter wings of Pallid can make it look more compact. Pallid's longer tail, narrower, more pointed wings and more buoyant flight distinguish it from Hen. Male distinguished by characteristic black wedge on primaries and much whiter head-to-breast than in other male harriers; above it has almost similar pale ash-grey shade as male Hen but primary pattern, paler head and lack of clear-cut white patch on uppertail-coverts distinguish Pallid. For distinction from male Montagu's see that species. Female and juvenile distinguished from similar Hen by flight, length and shape of wings and tail, more pronounced whitish collar and darker patch on ear-coverts. Juvenile is unstreaked on more rufous or orange underparts and underwing-coverts (streaked and less rufous in juvenile Hen). For distinction between female, juvenile and second-year male Pallid and Montagu's see latter species. Sub-adult male Pallid (second plumage: no juvenile flight- and tail feathers) in its dirty, dark, ash-grey (even brownish) plumage above, with only diffusely contrasting darker primary pattern, distinguished from same age male Montagu's by less black on primaries above, while black wedge is distinct below, lack of black secondary bands below and with diffusely banded tail-feathers; also by generally much paler head and upperbreast, white cheeks and less obvious, softer, more inconspicuous, long streaks on lower breast-belly. (In Montagu's head and breast darker grey in contrast with remaining distinctly rufous-streaked, white underparts; cheeks largely grey.)

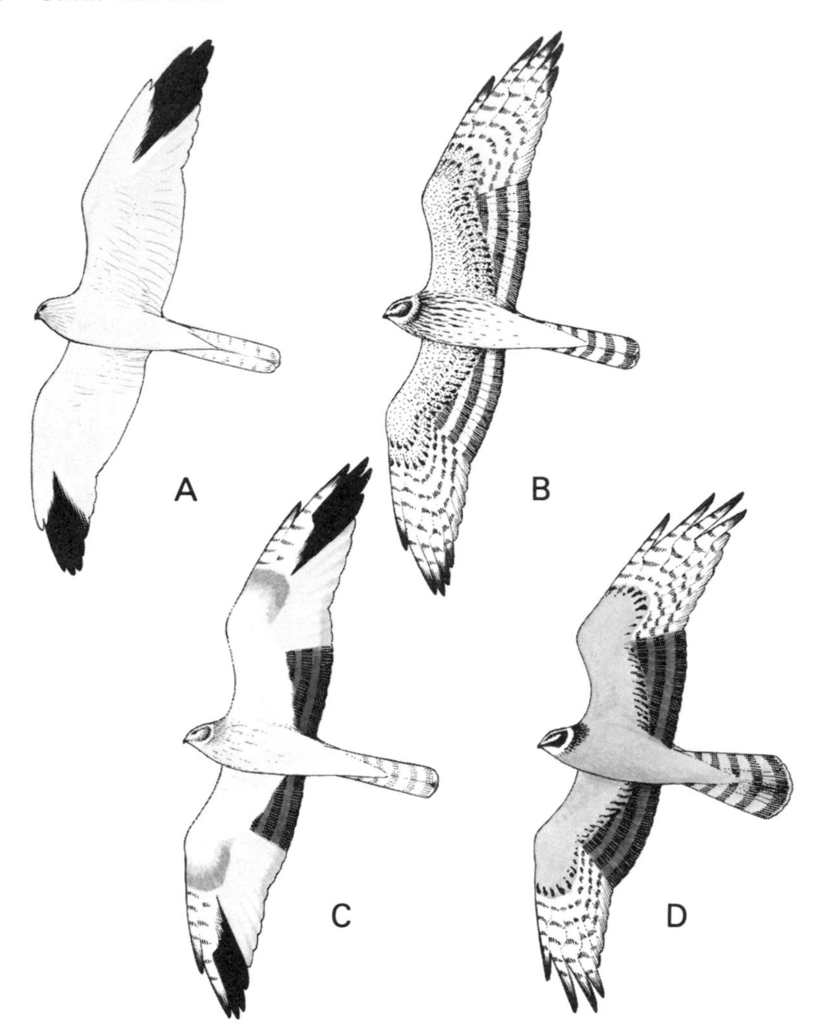

Fig. 41. **Pallid Harrier** *Circus macrourus*. Adult male, juvenile, transitional male and female from below. Male (A) is white apart from black wedge on 5th/6th primaries and barely discernible 7–9 greyish bars on tail and, at close range, a faint greyish wash on chin and throat and a diffuse pale greyish trailing-edge to wings. Adult female (B) resembles female Montagu's, though spacing between dark secondary bands is narrower and there is a whitish collar behind the dark crescent on ear-coverts. Underparts beige to pale rusty-yellow, densely streaked on throat-upperbreast, otherwise finely streaked. Juvenile (D) is very similar to juvenile Montagu's, having warm chestnut to yellowish-orange (even warm creamy) unstreaked underparts and wing-coverts and similarly barred tail and flight-feathers; dark-banded pale or dark grey secondaries below contrasting more or less with rest of underwing; individuals with yellowish-orange to warm cream seen more frequently among juvenile Pallid than Montagu's but there is an overlap in colour between two species. Head pattern shows more prominent black eye-streak, whitish collar behind blackish crescent on ear-coverts, which clearly reaches bill, and is generally darker on sides of neck behind pale collar which sometimes extends across throat (see Fig. 33). Second-autumn male (1+ years old), an example of which is shown in C, still shows juvenile outer primaries and inner secondaries while rest of flight-feathers, tail and coverts are in second (adult-like) plumage. Note white cheeks, remains of dark crescent on ear-coverts and diffuse pale collar; throat and breast are rusty- or creamy-white, usually with faint rufous streaks on breast and white belly. During second winter and following spring ($1\frac{1}{2}$–$1\frac{3}{4}$ years old) remaining juvenile flight-feathers moulted and second plumage attained.

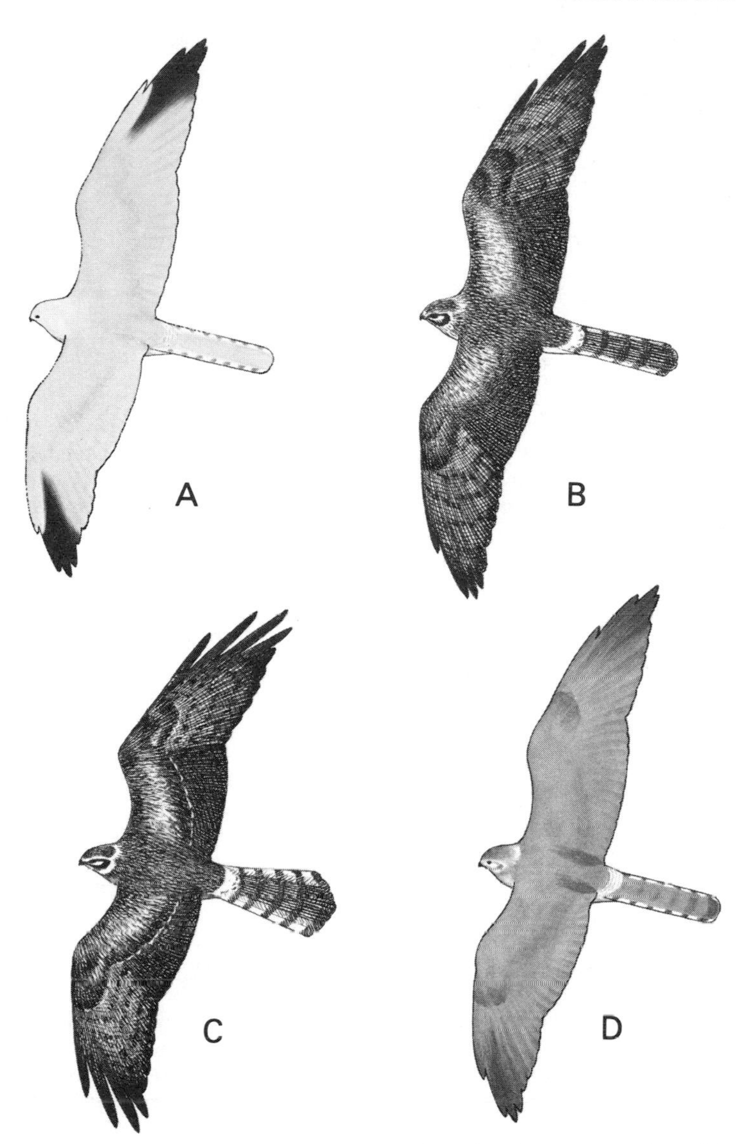

Fig. 42. **Pallid Harrier** *Circus macrourus*. Adult male, juvenile, adult female and transitional male from above. Adult male (A) is pale ash-grey above like male Hen, but slightly paler on back, clearly paler head to neck and only narrow black wedges on primaries; slightly darker barring on outer tail, while uppertail-coverts do not show up as a white patch. Adult female (B) is grey-brown above with creamy-buff patch on median to lesser coverts, smaller on average than in female Montagu's, a narrow white patch on uppertail-coverts and grey-brown central tail-feathers with about five darker bands; well-marked head pattern comprises whitish collar behind a dark crescent on ear-coverts which reaches base of bill, and a well-defined black eye-streak. Pale spot on nape often present and top of head usually darker than in female Montagu's, causing capped appearance. Juvenile (C) resembles female, but head pattern and whitish collar even more distinct. Upperparts darker with pale line along tips of greater coverts and well-developed whitish nape-spot. The third-year male (about two years old) depicted in (D) is considerably darker, browner and dirtier above than adult male with darker blackish-brown or grey primary wedge, less obvious than in adult; crown often darker and remains of dark crescent on ear-coverts, pale nape-spot and faint collar can be seen; faint barring on tail, and finely grey-barred, whitish uppertail-coverts.

Red Kite
Milvus milvus
(plates 54, 55)

Two Red Kites (left) with Black Kite

SILHOUETTE Wing-span 175–195 cm. Larger than Buzzard *Buteo buteo*, though of similar body length. Wings long, fairly broad and set well forward on rather slim body, an effect accentuated by the long tail. Tail shows a slight notch even when fully spread and a deep fork when closed. In head-on profile when soaring, wings very slightly arched.

FLIGHT In soaring, which it does frequently and magnificently, wings are held slightly forward from the body and often kinked with the carpal joint pressed forward. Wings constantly manoeuvred independently of each other and tail flexed from side to side with tips likewise moving independently. Flight easy, graceful and at times somewhat tern-like, the body rising and falling with the wing beats which are fairly deep, elastic and slower than those of Buzzard.

IDENTIFICATION Likely to be confused only with Black Kite. Red Kite is slightly larger but slimmer, lacking Black's more thick-set appearance, but this may not always be obvious unless the two are seen together. Otherwise, apart from Red Kite's more graceful movements, best distinction is shape, length and colour of tail: whereas Black's shorter tail is straight-ended when spread, Red's longer tail always shows at least a faint notch and normally a deep fork; Red Kite's tail is also much redder, particularly above (though less rusty in juvenile) than that of even the most brightly marked (young) Black Kite and from below and against the light is a rich transluscent orange (at best dull orange in the case of the Black Kite). At the same time Red Kite's whole plumage is paler and brighter, being a mixture of red-browns instead of sombre browns; the head is whitish, nearly always paler than that of Black, and the pale band across the upper-wing coverts is broader and more conspicuous, and primaries below show larger whiter and much less barred patch than in any European Black Kite.

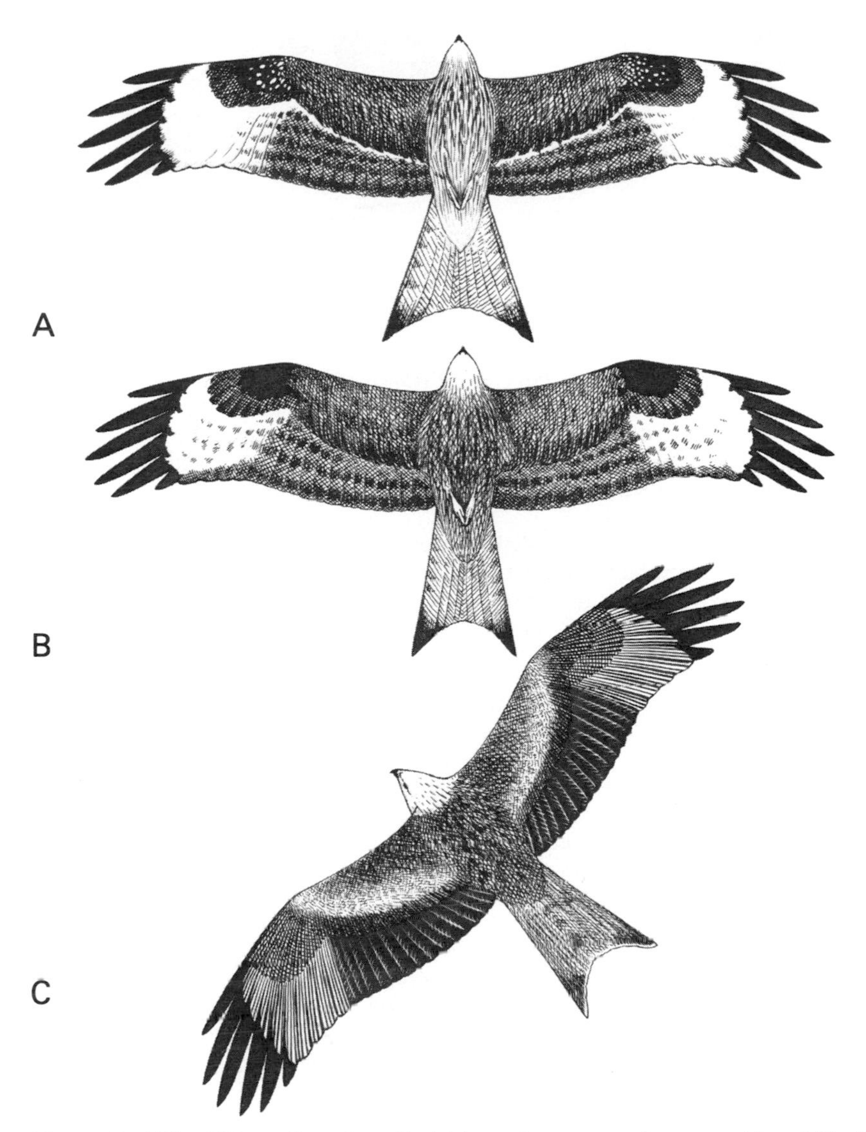

Fig. 43. **Red Kite** *Milvus milvus*. Juvenile (A) has pale creamy under-parts with variable dark spots or streaks on lower neck and breast, the unspotted pale belly-undertail-coverts of similar pale shade as undertail, which has a diffuse darker subterminal band and blackish outer tips. Pale underparts contrast somewhat to darker underwings; dark secondaries, black wing-tip and a large white patch on primaries nearly filling the width of each wing; a pale line at tips of greater coverts more or less conspicuous. Adult from below (B) has darker, more reddish-chestnut body and lesser-median underwing-coverts with, particularly on breast, broad black shaft-streaks; less or no contrast between body and underwing-coverts, but contrast between dark undertail-coverts and paler undertail which lacks diffuse subterminal band, but has blackish outer tips; usually lacks pale line on greater coverts of juvenile. Adult from above (C) has whitish crown, rather darker reddish-chestnut body and forewings, dark brown flight feathers though slightly paler diffuse patch on primaries and black wing-tips; an often conspicuous buffish band across wing coverts. Uppertail bright rusty-red with darker outer tips. Juvenile above (not illustrated) differs from adult in having a pale line on tips of dark greater coverts, generally broader and paler band on coverts and less rusty uppertail with diffuse dark subterminal band (visible on spread tail); head also darker than adult but with whitish streaks around neck.

Black Kite
Milvus migrans
(plates 56, 57)

SILHOUETTE Wing-span 160–180 cm. A little smaller and more compact than Red Kite (compare Figs. 34A and B). Wings and body have coarser and more thick-set appearance; from side wings look well bowed forward. Tail shorter and less deeply forked, becoming almost straight-ended with triangular shape when spread. In head-on profile when soaring or gliding, wings slightly arched.

FLIGHT In soaring or gliding, slightly arched wings are rarely held rigid, but constantly flexed giving buoyant effect; tail also manoeuvred like Red Kite's. In active flight, wing beats are slower and more floppy than those of Buzzard *Buteo buteo* and interspersed with glides; head often pointed downward as bird peers from side to side, combining with the slightly arched wings to give hunched look.

IDENTIFICATION Possibility of confusion, particularly of well-marked immature, with Red Kite and caution needed in identifying young kites on autumn passage. Black more thick-set and less graceful with duller, shorter and less forked tail which is never rusty-red; pale patch on primaries below not so large, white and sparsely barred. (Note too that Black start to move south in August and most have left Europe by end September, while October is main passage month for Red; in breeding season Black often frequent town edges, rubbish dumps and water, whereas Red are found more in open, lightly wooded country, but there is considerable overlap.) May also be confused with all-brown Marsh Harrier *Circus aeruginosus,* but flight very different: Black Kite soars and glides on slightly arched wings and twists tail constantly from side to side, while Marsh Harrier always has wings raised in shallow V. Dark-phased Booted Eagle can sometimes be confused, but Black Kite has tail forked when closed and straight-ended when spread, while Booted has tail straight-ended when closed and rounded when spread; other distinctions under Booted.

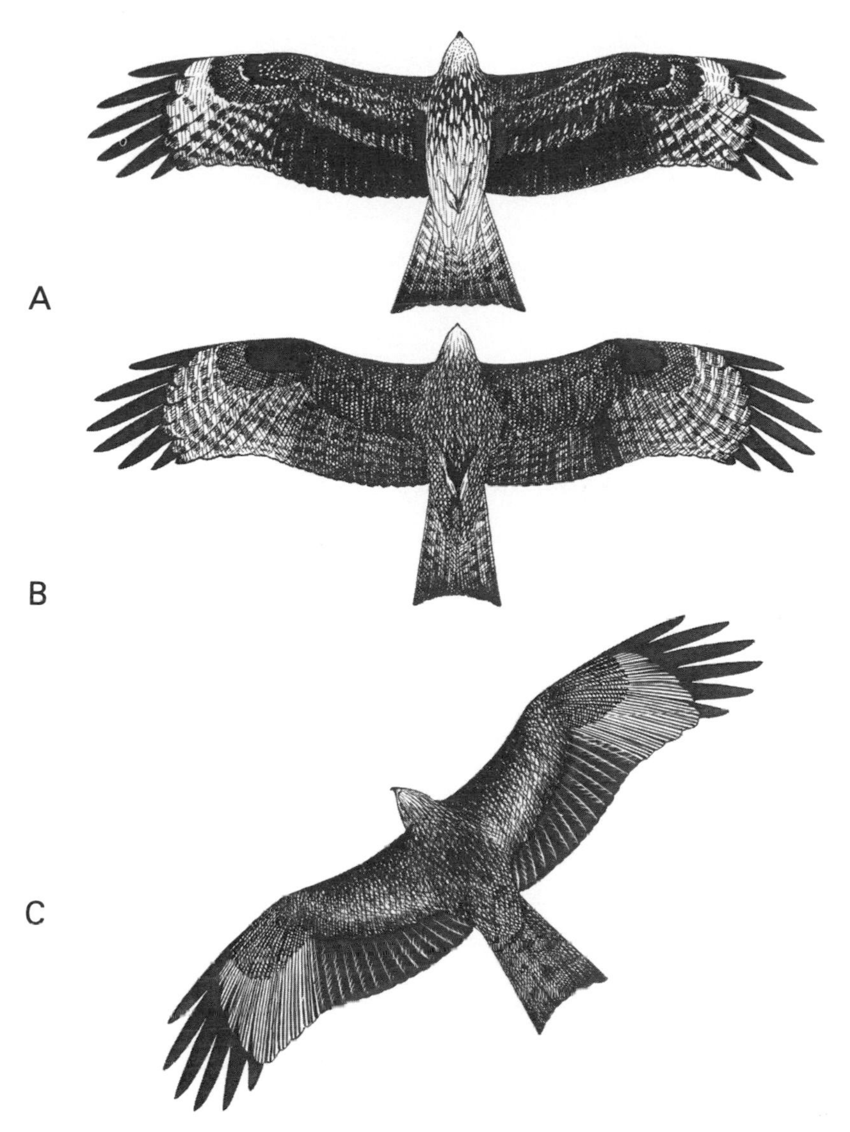

Fig. 44. **Black Kite** *Milvus migrans*. Typical juvenile from below (A) has pale rufous-brown to buffish underparts with darker brownish spots and streaks on lower neck and breast; the generally pale underparts contrast with darker underwings which have pale area (variable in size) at base of primaries; pale line on tips of greater coverts. Undertail rather pale rusty-yellow at base with distinctly darker terminal half and broad creamy bar at tip; basal part of tail can appear transluscent rufous. Adult from below (B) has all underparts uniformly darker: dark rusty-brown, the lower belly-undertail coverts not contrasting with dark underwings; usually lacks pale line on greater coverts; undertail not bicoloured as in juvenile but generally pale with dark barring and with no creamy terminal bar. Primaries below usually paler than secondaries but palish patch can be almost absent. In certain lights the translucent tail appears paler than undertail-coverts. Adult from above (C) is medium-brown with paler head and dark brown flight feathers; diffuse pale patch on primaries; wing coverts pale rusty to buffish-brown forming pale band across inner wing, generally less conspicuous than in Red Kite. The juvenile from above (not illustrated) differs from adult in having whole upperparts and wing-coverts spotted and streaked buffish, a pale line on tips of greater coverts and a dark patch from eye to ear-coverts more pronounced.

4 The Vultures

The four vultures in our area are restricted to countries bordering the Mediterranean and Black Seas. The group is unique among European raptors in feeding almost entirely on carrion: Griffon *Gyps fulvus,* Black *Aegypius monachus* and Egyptian *Neophron percnopterus* flock, sometimes together, at animal carcases or, particularly in the case of the Egyptian, on rubbish dumps, while the more solitary Lammergeier *Gypaetus barbatus* sails over high crags, searching for bones. The Griffon, Black and Lammergeier are the largest raptors in Europe, with wing spans of eight to nine feet (245 to 275 cm); the Egyptian is much smaller. The huge wings and short tails of both Black and Griffon can give an eagle-like appearance; but there should be few real problems with the long diamond-shaped tail and pointed wings of the Lammergeier, or with the fairly broad wings, wedge-shaped tail and black and white plumage of the adult Egyptian (although brown immatures may cause difficulty).

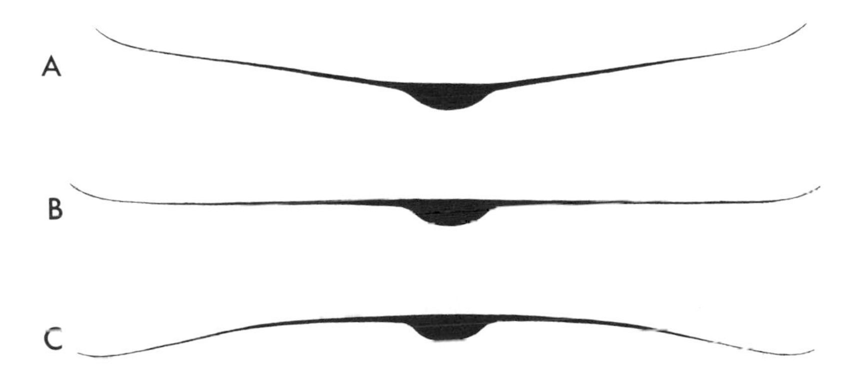

Fig. 45. Head-on profiles of vultures: (A) soaring Griffon *Gyps fulvus*; (B) soaring Black *Aegypius monachus* (Lammergeier *Gypaetus barbatus* and Egyptian *Neophron percnopterus* often adopt this position too, though both sometimes also soar with wings bowed at the carpal joints); and (C) frequent gliding position of all four.

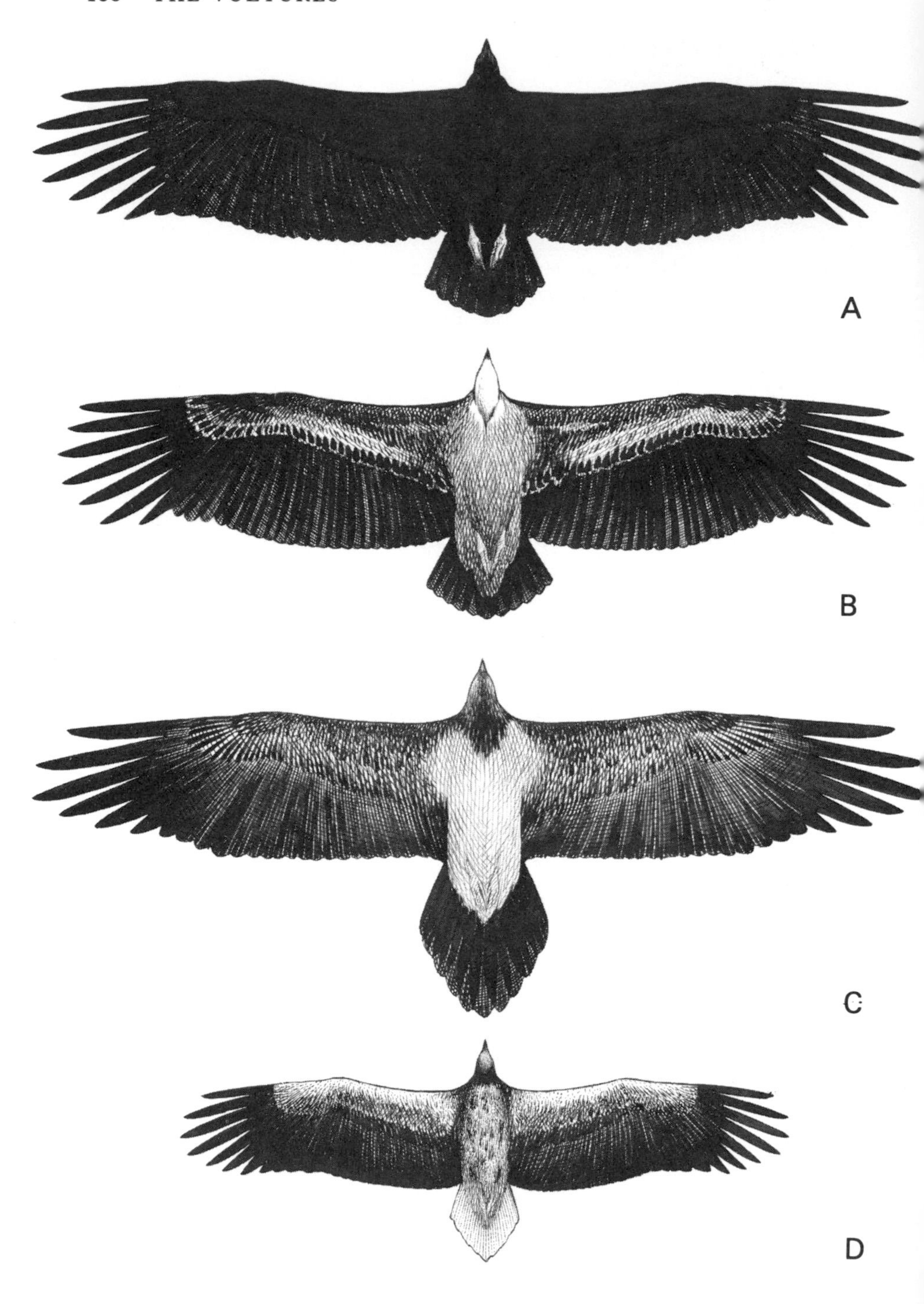

Fig. 46. Undersides of the four vultures.

Fig. 46A. **Black Vulture** *Aegypius monachus*. Juvenile shown. Huge, with long, broad wings, short head and tail; proportions similar to Griffon, but wings more even in width, lacking marked bulge to rear edge; at distance can resemble huge Spotted Eagle *Aquila clanga*. Plumage never contrasted as Griffon, but all blackish, even in faded old birds, with black forewings and body darkest parts; pale feet often conspicuous overhead. Unlike Griffon, soars on flat wings (45B). Although normally singly or in twos and threes, and generally scarce, parties up to 15 not uncommon in some areas; largely resident in both open highlands and lowlands, pairs usually nesting singly in trees, in Iberia, Balearic Islands, Sardinia, Sicily (?), south Balkans north to Romania, south Russia, Turkey, Cyprus, Caucasus, Middle East, and Morocco (?).

Fig. 46B. **Griffon Vulture** *Gyps fulvus*. Adult shown. Proportions similar to Black, but very short innermost primaries produce marked bulge to rear edge of secondaries. Contrast of buffish body and forewings with blackish primaries and tail provides further distinction. Unlike Black, soars with wings in shallow V (45A). Much commoner and more gregarious than Black, frequently in groups of up to 60; partly migratory in similar habitats, but nesting in colonies on cliffs (both caves and ledges), in Iberia, south France, Sardinia, Italy, Sicily, Balkans north to Romania, Turkey, Cyprus, Caucasus, Middle East and Morocco, Algeria, south Tunisia and Egypt.

Fig. 46C. **Lammergeier** *Gypaetus barbatus*. Juvenile shown. Size similar to Griffon, but rather falcon-like outline tends to give smaller appearance unless another species present for comparison; distinctive shape quite unlike any other European raptor, with long broad wings, tapering hands, and long, ample, diamond-shaped tail. Juvenile and other immature stages all dark sooty-brown except for paler, brownish-grey underbody; adult has bright creamy or orange-yellow underbody and yellowish head (with broad black stripe through eye) contrasting with blackish wings and tail. Soars and glides on flat or slightly bowed wings. Usually singly, in pairs or in family parties, and scarce and apparently rapidly decreasing in most areas; largely resident in high mountain country (occasionally foothills), in Spain, France (Pyrenees), Corsica, Sardinia, south Yugoslavia, Greece, Bulgaria, Turkey, Caucasus, Middle East, and Morocco, Algeria and Tunisia.

Fig. 46D. **Egyptian Vulture** *Neophron percnopterus*. Third-autumn shown. Smallest vulture, with narrow pointed head accentuated by long bill, fairly long wedge-shaped tail, and long, fairly broad wings of even width except for tapering hands. Juvenile plumage all sooty-brown with little contrast apart from pale rump and sometimes pale bars on wing-coverts; later, immature paler on back and uppertail-coverts, upper secondaries and underwing-coverts and dark ruff retained to sub-adult plumage; adult has black flight-feathers (with white centres to upper secondaries) contrasting with white body, tail and wing-coverts, though often dark smudges on coverts and head dirty yellowish. Soars and glides on flat or slightly bowed wings. Widespread and often gregarious, particularly around rubbish dumps, but pairs nest singly; summer visitor to all kinds of open country from mountains to lowland village areas, in Iberia, Balearic Islands, south France, Italy, Sicily, Balkans, Turkey, Caucasus, Middle East, and across north Africa.

Black Vulture
Aegypius monachus
(plates 58, 59)

Black (left) and Griffon Vulture (right)—going away

SILHOUETTE Wing-span 250–295 cm. Very large vulture with huge wings (compare Figs 46A with 46B–D), much larger than any eagle. Head protruding little. Wings long and broad, but without bulging secondaries of Griffon, so that front and rear edges are more nearly parallel. Tail very short and, although slightly longer and more wedge-shaped in fresh plumage than that of Griffon, not noticeably different in the field. In head-on profile when soaring, wings held level; when gliding, wings bowed at carpal joints.

FLIGHT In soaring, wings are invariably held flat with just tips of primaries upturned; spends much time soaring effortlessly with hardly a wing beat. In gliding, arms are held straight out from body and hands bowed at carpal joints, which are pressed slightly forward with primaries closed and angled back; wings are held motionless with only occasional downward flap. When launching off from ground or cliff face, wing beats are slow and deep.

IDENTIFICATION Confusion likely only with Griffon which, however, soars with wings in shallow V, has more bulging secondaries and shows more plumage contrast. Black soars on flat, straighter-edged wings and is mainly sooty-black, darkest on coverts and body, though old individuals become slightly paler with light sooty-brown areas along edge of underwing-coverts. Always much darker than Griffon (even though latter can sometimes look dark in certain lights) and at distance often appears jet black, not unlike huge adult Spotted Eagle *Aquila clanga* in plumage. Head pale in older birds, brownish in juvenile, never as white as in Griffon; brownish ruff inconspicuous in field. Pale feet, which show up surprisingly at closer ranges, are diagnostic. Immature White-tailed Eagle *Haliaeetus albicilla* is also dark with similarly shaped wings and short tail, but, even if whitish centres to tail-feathers cannot be seen, has markedly protruding head and neck. Adult Imperial Eagle *Aquila heliaca* can appear very like Black Vulture, particularly when soaring at a distance, but has wings proportionately narrower, tail longer and fuller and head more protruding; at closer ranges, Imperial's bicoloured tail, usually yellow vent, undertail coverts and crown to hind-neck and white 'braces' diagnostic.

Fig. 47. (*Opposite*). **Black Vulture** *Aegypius monachus*. Juvenile (A) and adult (B and C) are similar, though juvenile is blacker and typical individuals are almost entirely sooty-black (blackish-brown at close ranges) with the wing-coverts and body slightly darker than primaries and secondaries; in adults a pale head with blackish eye-mask (head all dark brown in juveniles) protrudes from a brownish ruff and pale feet, ranging in colour from bluish-grey and bluish-white to pale yellow, show up well against the blackish undertail-coverts. Adults and sub-adults are generally paler and browner with a wide band of light brownish along the underwing-coverts, or occasionally more than one paler band.

Fig. 47. Black Vultures from below and above.

Griffon Vulture

Gyps fulvus
(plates 60–63)

SILHOUETTE Wing-span 243–270 cm. Almost as large as Black Vulture (compare Figs. 46B with 46A). Head again protruding little and, though slightly narrower than that of Black, shape too similar to be valid as field-character. Wings long and broad, but innermost primaries shorter than those of Black, producing pinched effect which gives a marked bulge to the secondaries and thus a deeply curved rear edge to each wing. Tail very short and similar to that of Black; it is, in fact, more square-cut though where abraded appears wedge-shaped or rounded. In head-on profile when soaring, wings raised in shallow V, quite different from soaring positions of Black and other vultures; when gliding, wings bowed at carpal joints.

FLIGHT In soaring, wings are raised in shallow V; like other vultures, spends much time soaring effortlessly with hardly a wing beat. In gliding, as Black Vulture, arms are held straight out from body and hands bowed at carpal joints, which are pressed slightly forward with primaries closed and angled back; wings are held motionless with only occasional downward flap. When launching off from ground or cliff face, wing beats are slow and deep.

IDENTIFICATION Likely to be confused only with Black Vulture, but wing position and shape when soaring and contrast in plumage provide good distinctions. Griffon soars on raised wings with bulging secondaries, whereas Black soars on flat wings with front and rear edges less bulging more nearly parallel. From below, juvenile shows marked contrast of gingery-buff underbody and underwing-coverts with black flight-feathers and tail; contrast becomes less striking as bird grows older, but even full adult's body and coverts are never as dark as those of palest Black Vulture. Greyish feet do not show up like pale ones of Black Vulture, but very pale creamy-white head and neck, though often darkened by blood and filth, are usually conspicuous, as is white ruff of adult. From above, buffish-grey to brown coverts and rump contrast with blackish flight-feathers and tail, unlike entirely black-brown upperparts of Black Vulture.

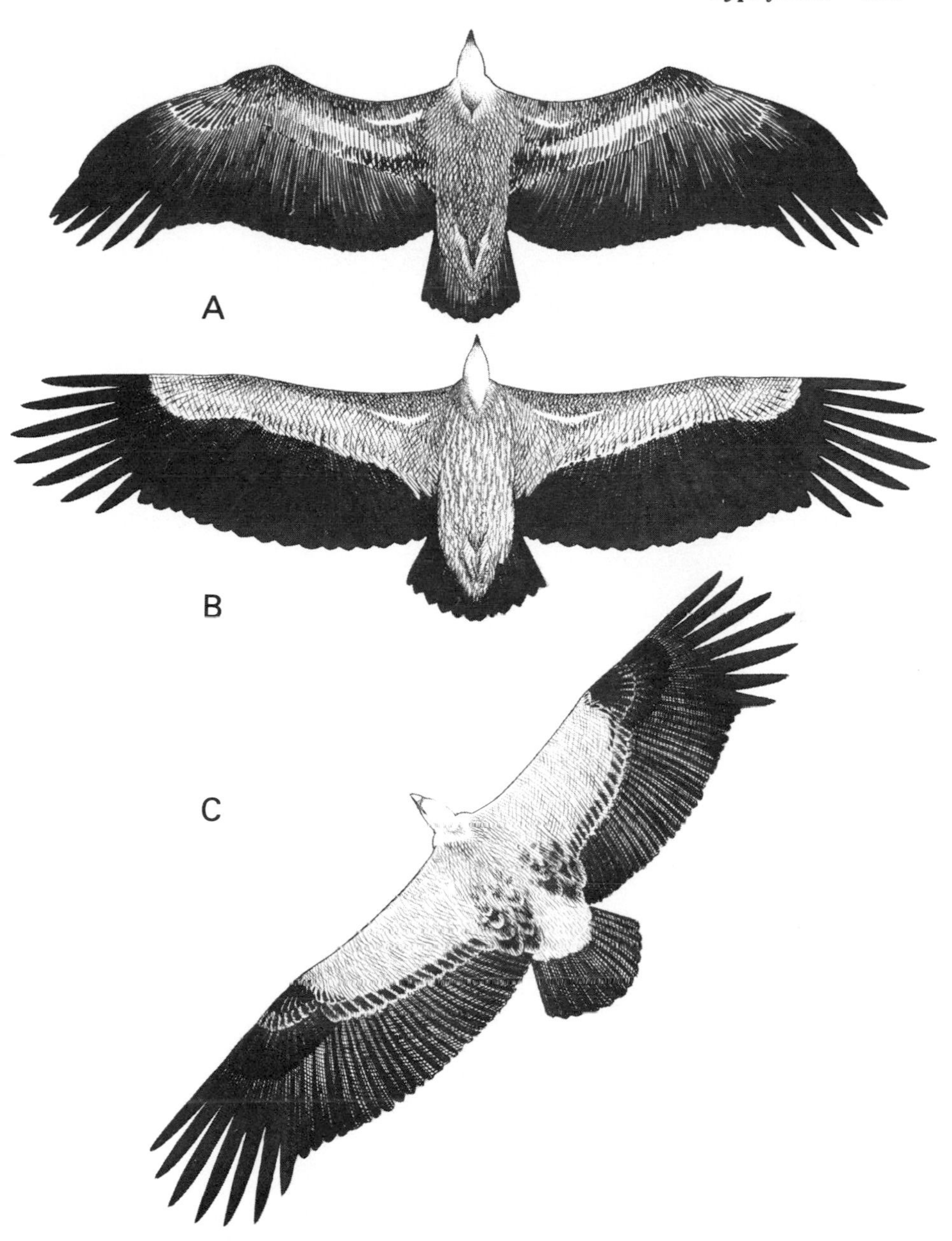

Fig. 48. **Griffon Vulture** *Gyps fulvus*. Adult and juvenile from below and adult from above. Adult Griffon (A and C) has gingery-buff underwing-coverts and underbody contrasting with blackish primaries, secondaries and tail; pale (often almost whitish) lines and bars on the lesser and median coverts increase contrast which can usually be seen even in poor light or at great distances; feet are greyish and so not conspicuous; head and neck are off-white (but may be stained darker after feeding). Juvenile (B) has much sandier underwing-coverts and underbody, latter profusely streaked off-white, a feature only noticeable at close range, though helping to make underbody paler than adult's, and contrast with the blackish secondaries, primaries and tail is even more striking. Adult from above (C) has a whitish head and ruff and gingery-buff upperwing-coverts, back and rump contrasting with black flight-feathers and tail. Juvenile has duller, buff-brown ruff.

Lammergeier

Gypaetus barbatus
(plates 64, 65)

SILHOUETTE Wing-span 250–280 cm. Very large, with wing span similar to or even greater than Black Vulture. Head small, but quite well protruding. Long, narrow and rather pointed wings and long, ample, diamond-shaped tail produce an outline unique among European raptors. Juvenile has longer secondaries than adult and, especially when tail abraded, tends to have less distinctive shape, not unlike huge immature Egyptian Vulture. In head-on profile when soaring and gliding, wings either level or with hands slightly lowered.

FLIGHT In soaring and gliding, wings are held flat, or with slightly lowered hands, and straight out from body; soars for long periods over ridges and precipices, or back and forth in front of cliff face. In gliding, as with all vultures, arms are held straight out from body and hands bowed at carpal joints, which are pressed slightly forward with primaries closed and angled back. When taking off, wing beats are heavy and ponderous; otherwise, virtually no flapping except for occasional down beat.

IDENTIFICATION Hardly possible to mistake for any other bird. Adult's orange-buff or yellowish-buff underparts, in contrast to dark underwings, are conspicuous at great ranges, as is the yellowish head which contrasts strongly with the slate-black upperparts when seen from above or while banking. When no other birds present for size comparison, a heavily abraded immature can look almost as short-tailed and nearly as dark as young Egyptian, but wings are clearly longer and comparatively narrower, and wing-tip more pointed.

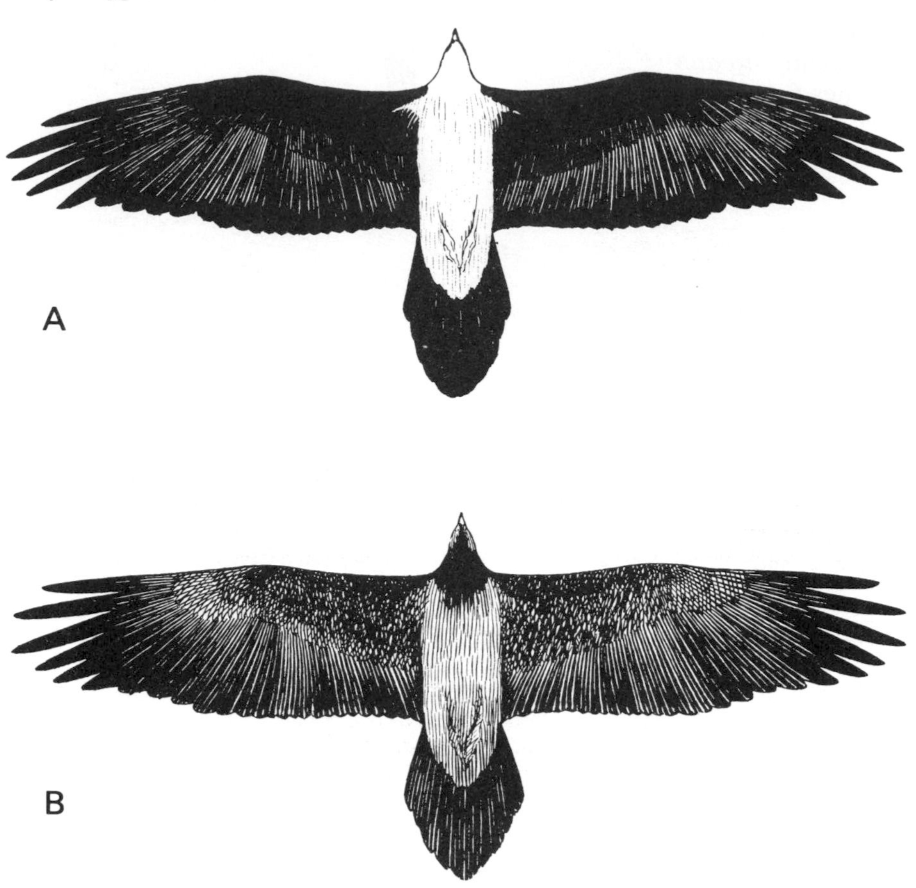

Fig. 49. **Lammergeier** *Gypaetus barbatus*. Adult and juvenile from below. Adult (A) has bright creamy-white or yellow head shading to darker rufous-cream, somewhat orange, on the underbody and undertail-coverts; a dark, usually incomplete pectoral band of small black spots and darker rufous on throat is often apparent at close range; lesser and median underwing-coverts are black; greater coverts and flight-feathers are dark slate-grey shading to blackish ill-defined band along trailing edge of wings; in good light there is clear contrast between flight-feathers and blackish forewing, but otherwise underwings look all dark in contrast to paler underparts; tail slate. Juvenile (B) has blackish-brown head and neck and dirty brownish-grey body; flight-feathers as in adult, though tinged brownish, not slate; underwing-coverts, which have a scruffy appearance, are grey-brown; partly grown moustaches are not easily seen even at close range.

Fig. 50. Adult and juvenile Lammergeiers *Gypaetus barbatus* from above. Adult (A) at a distance looks simply blackish with a brilliant creamy-yellow head. Closer views reveal a slate or charcoal grey plumage with (when very close) white shaft-streaks; striking black stripe from bill through each eye and black feathers drooping below the bill. The juvenile (B) is dark grey-brown above, darkest on flight-feathers and tail, the latter being tipped dull buff; the head and neck are mainly black-brown, with dingy buff areas around eye, on crown and upper nape; there is a whitish-buff patch on mantle, flecked with dark brown, and wholly pale buff and buff-edged feathers on otherwise dark grey-brown wing-coverts, forming, most noticeably, a light, ragged bar on median coverts; the moustaches are only partly grown at this stage and hardly noticeable. With increasing age, head and neck become paler while mantle darker and adult plumage is attained by the age of 4–5 years.

Egyptian Vulture
Neophron percnopterus
(plates 66, 67)

SILHOUETTE Wing-span 160–170 cm. Much smaller and more lightly built than other vultures (compare Figs. 46D with 46A–C); size between Black Kite *Milvus migrans* and Golden Eagle *Aquila chrysaetos*, approximating to that of Lesser Spotted Eagle *Aquila pomarina*. Head small, but projecting well to distinct point owing to long, delicate bill (see also plates). Wings long and fairly broad with parallel edges but tapering hands. Tail wedge-shaped and medium-short in length, shorter than width of wings; adult has broader base to tail than immature because of extra feathers at sides. In head-on profile when soaring, wings either level or bowed with drooping hands; when gliding, wings bowed at carpal joints.

FLIGHT In soaring, wings either held flat or angled down from carpal joints and there is much more flexibility than in other vultures. In gliding, wings bowed at carpal joints, with primaries angled back, as with the other species. Active flight includes more wing beats than larger vultures, though much time still spent soaring: takes off with heavy and quite deep wing beats and moves surprisingly fast when disturbed from rubbish dump or carcase.

IDENTIFICATION Adult not likely to be confused with any other raptor, although at a quick glance can resemble light-phase Booted Eagle *Hieraaetus pennatus*, the only other European bird of prey with similarly patterned black and white underparts. (White Stork *Ciconia ciconia* also has similar flight pattern, which has been known to cause confusion, but long neck and legs protrude well out in front and behind.) Brownish immature superficially resembles any other large brown raptor, but more pointed and comparatively narrow (though still fairly broad) wings, wedge-shaped tail (which often appears translucent orange-buff) and pointed head should readily distinguish it. Young Lammergeier, which has slightly shorter tail and longer secondaries than adult, can resemble immature Egyptian but is much larger and comparitively longer-tailed and has more contrast between all dark head-upperbreast and rest of underparts, and has darker tail than Egyptian.

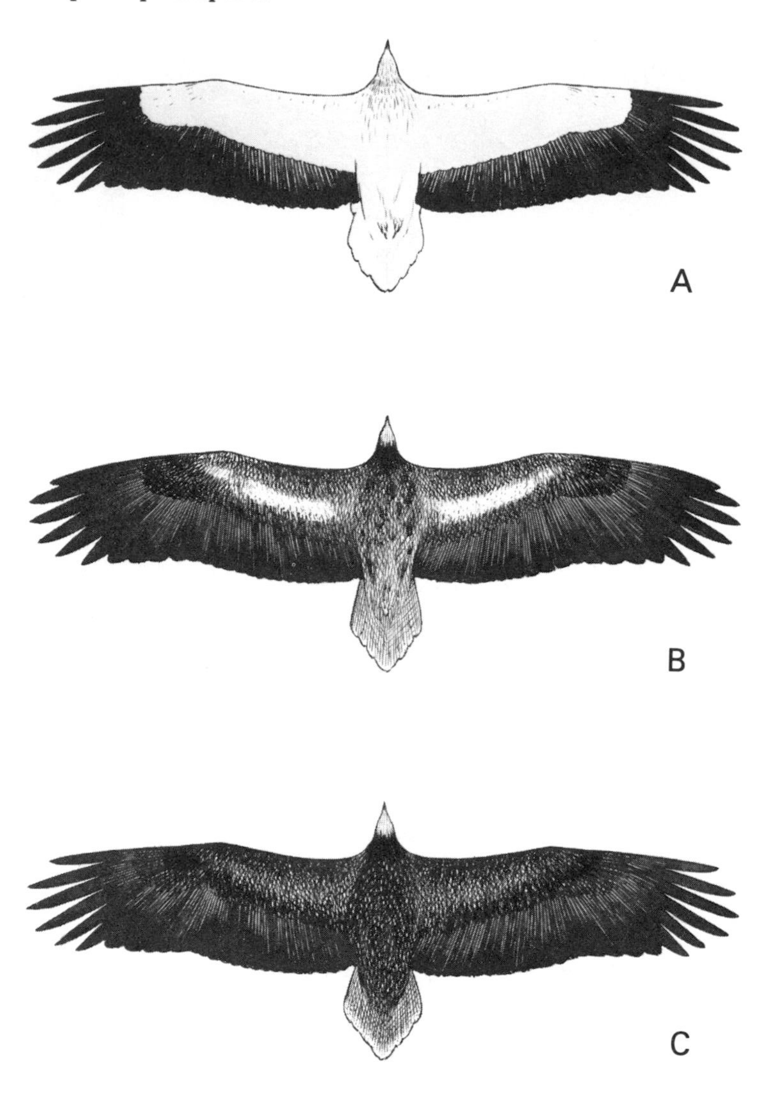

Fig. 51. **Egyptian Vulture** *Neophron percnopterus*. Adult, immature and juvenile from below. Adult (A) is distinctive, having white body, wing-coverts and tail contrasting with blackish primaries and secondaries; there is a certain amount of dirty yellowish discolouration around the neck and upperbreast and some individuals, probably those moulting from sub-adult to adult plumage, have dirty smudges on the white coverts and body; against a strong light the tail is translucent and often shows an orange hue. Juvenile (C) is wholly blackish-brown though sometimes paler and browner across median coverts, or with all lesser and median coverts dark brown, finely tipped paler to give coverts a dark grey-brown appearance; chin and throat are dirty cream, but ruff blackish-brown, frequently darker than the dark brown underparts. Tail is medium brown with a diffuse whitish tip, which may wear off; tail can take on an orange tinge when the bird is soaring in strong sunlight. Immature (B) is variable but with increasing age lighter areas appear on the wing coverts, belly and tail-coverts; tail also lightens in colour becoming more buff-brown or pale greyish. From this stage more and more light feathers appear until the adult plumage is attained and the variety of immature and sub-adult plumages is thus considerable. As a rule the ruff remains very dark even up to the sub-adult plumage.

Fig. 52. **Egyptian Vulture** *Neophron percnopterus*. Adult, immature and juvenile from above. Adult (A) is white on head, forewings, back, rump and tail; flight-feathers are black with creamy or silvery-grey outer webs which make the secondaries appear white-streaked, though the bases are black, often forming a bar; primary coverts and innermost greater coverts black. The white parts are often tinged yellowish, rusty or dirty greyish, particularly on nape, hind neck, scapulars and median coverts. Bare skin around the eyes and base of bill is bright yellow, but only noticeable at close range. Juvenile (C) has blackish-brown ruff, dark brown mantle and wing-coverts, the latter with pale creamy buff tips forming wing-bars most noticeable on median coverts; smaller scapulars brownish with creamy tips and edges, forming dark brown smudgy area bordering vague pale 'braces'; remaining scapulars pale brown fading into creamy centre of lower back and rump, the latter often mottled brown, thus contrasting with whitish uppertail-coverts. Tail grey-brown with ill-defined pale terminal bar. Blackish primaries with pale brown outer webs; dark brown secondaries with creamy inner webs and dark bases. Bare patch around eye is bluish-grey. Pale bars to wing-coverts and tip of tail worn off with time, the former then appearing blotched. Immature plumages variable depending on stage of moult, more creamy feathers appearing with time. (B) shows an immature about half-way to adult plumage (about two years old). Plumage is similar to adults but tail is grey-buff with paler tips and other light parts are dirty cream, even yellowish, with various brownish smudgy marks on back and wing-coverts; ruff is still blackish-brown contrasting with pale creamy-yellow mantle. Bare skin of the face gradually changes from juvenile bluish-grey to greyish-yellow in sub-adult, which also has scattered pale brown smudges on yellowish-white mantle and wing-coverts; its tail is not pure white and the ruff remains darkish.

5 The Large Falcons

The cosmopolitan genus *Falco* is one of the largest genera of raptors; in Europe it is represented by ten breeding species, the four biggest of which are dealt with in this group. These four, the Gyrfalcon *F. rusticolus*, Saker *F. cherrug*, Lanner *F. biarmicus* and Peregrine *F. peregrinus*, present problems of field identification generally, but especially in their juvenile plumages. This applies particularly to the Saker and Lanner. To some extent three of the four species replace each other geographically within the west Palearctic: the Gyrfalcon in Iceland, northern Fenno-Scandia and arctic Russia, the Saker in eastern Europe and west-central Asia, and the Lanner in south-east Europe, the Middle East and north Africa. The almost worldwide Peregrine overlaps all three, but has declined considerably in Europe during the last two decades.

All four are found in open country, from tundra through to moorland and desert. They are extremely conservative in their choice of breeding sites, which are usually on cliff ledges, and the same nests may be used for many years. Like other falcons they do not build nests of their own, but either take over old ones of other species or dispense with them entirely. They are all roughly the size of Woodpigeons *Columba palumbus* (though the Gyr is larger), with rather long, broad-based wings and medium-length, tapering tails. They are powerful hunters and have rather similar hunting techniques, being capable of tremendous stoops on live prey, though the Gyr rarely dives at its quarry in this way. The usual prey is birds, sometimes of very considerable size.

These falcons were once commonly used in falconry, and even today young are taken from nests and trained; most of these are Lanners in Arab countries, but the sport seems to have had a renaissance in Europe in recent years and is thought by many to be a threat to the declining European populations.

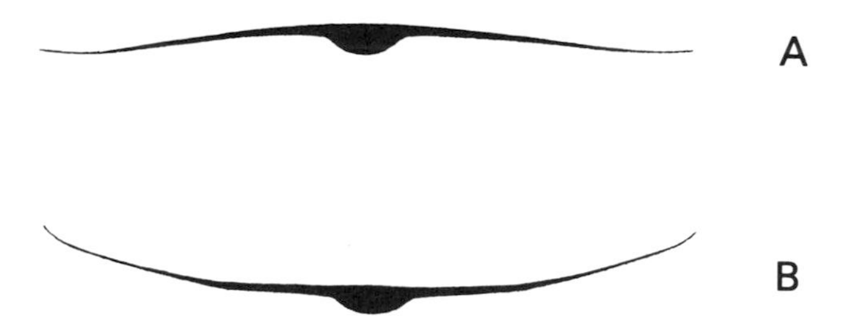

Fig. 53. Head-on profiles of Gyrfalcon *Falco rusticolus*, Saker *F. cherrug*, Lanner *F. biarmicus* and Peregrine *F. peregrinus*: (A) is the typical gliding and soaring position of all four, but occasionally a wing position (B) almost like that of harriers *Circus spp.* is adopted when either soaring or gliding, particularly by Gyr and Saker.

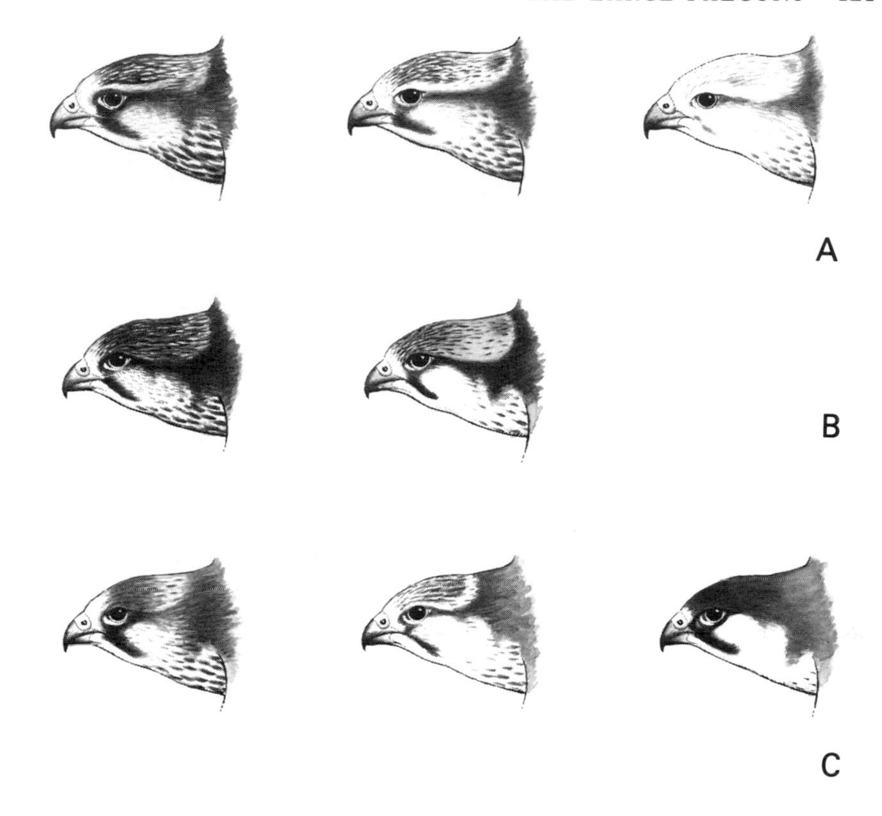

Fig. 54. Head patterns of some large falcons. A. **Saker** *Falco cherrug cyanopus*. Left, juvenile; centre and right, adults. B. **Lanner** *F. biarmicus feldeggi*. Left, juvenile; right, adult. C. **Peregrine** *F. peregrinus calidus*. Left and centre, juveniles; right, adult (pale type). For details see text.

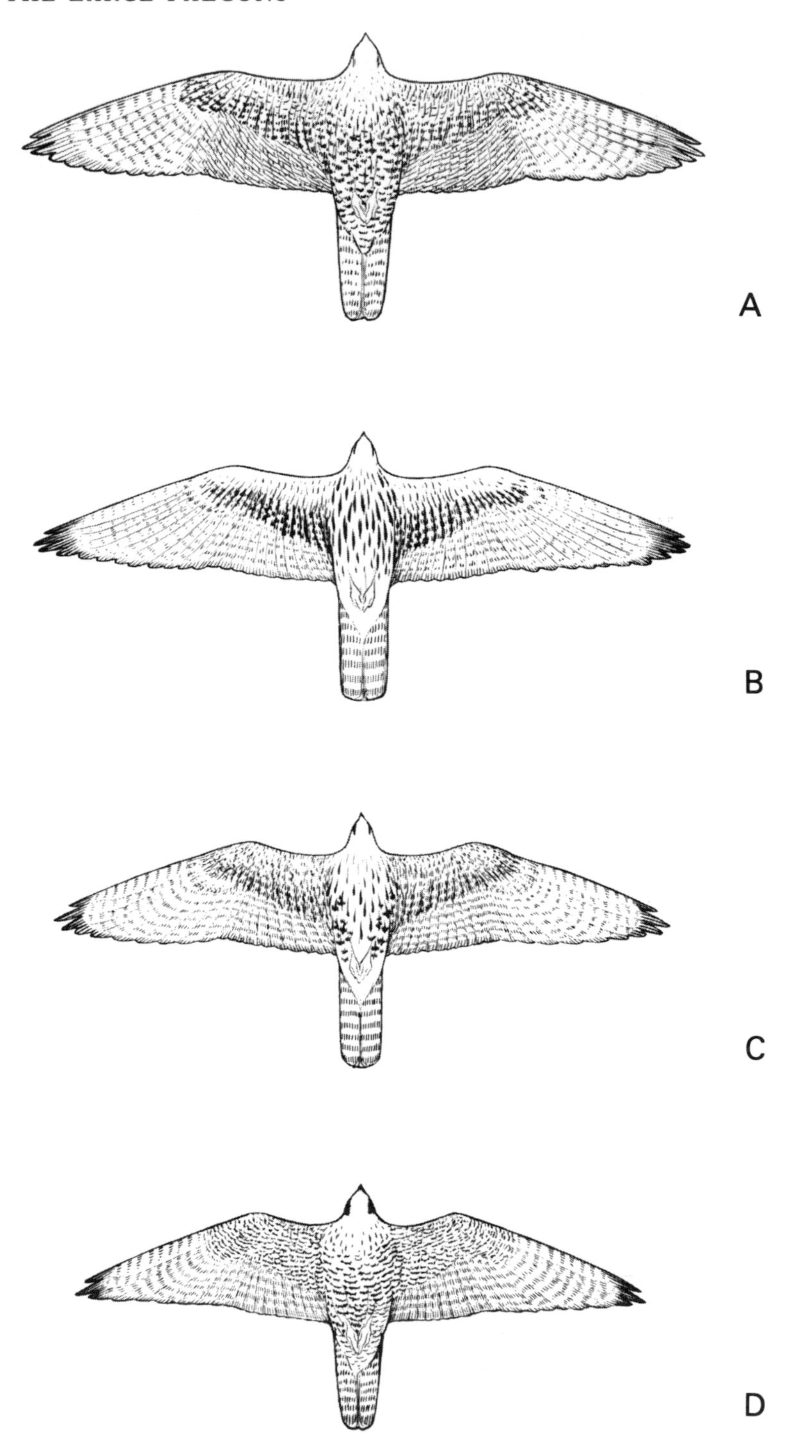

Fig. 55. Typical undersides of grey phase adult Gyrfalcon and of adult Saker, Lanner and Peregrine.

Fig. 55A. **Gyrfalcon** *Falco rusticolus*. Grey phase adult shown. Largest of the group, with heavy body, long and broad wings, particularly at base, and moderately long tail. Shows infinite variation in plumage from almost pure white phase ('*candicans*') through grey ('*islandus*') to relatively dark grey-brown forms ('*rusticolus*'). Moustachial stripe usually not very conspicuous, but may be more so on dark phase birds, especially juveniles. Circumpolar, usually in remote, rocky, open country and on sea coasts, sometimes near coniferous forests. Breeds Greenland (predominantly white phase, though some grey, especially in east and south) and Iceland (almost all grey), a few wandering erratically in winter to Ireland, Britain and Faeroes; and northern Fenno-Scandia and Russia (nearly all grey or dark), wintering irregularly south to southern Scandinavia and Baltic States and rarely (probably mostly young birds) to western and central Europe.

Fig. 55B. **Saker** *Falco cherrug*. Typical adult shown. Slightly smaller than Gyr but larger than Peregrine. Wings and tail relatively longer than Peregrine's; wings rather rounded at tips. Brown colour of upperparts, especially of adults, not unlike that of female Kestrel *F. tinnunculus*. though *saceroides* morph is dusty grey-brown. Underparts and underwings whitish streaked brownish on body and especially greater and median coverts. Moustachial stripe usually very inconspicuous. Found on open plains, steppes, semi-deserts, breeding eastern Europe from Czechoslovakia to east Balkans and Ukraine, and central Turkey. Partly migratory, especially northern populations and juveniles; in small numbers from September to April in south-east Europe, north-east Africa and Middle East.

Fig. 55C. **Lanner** *Falco biarmicus*. Typical adult shown. Size between Saker and Peregrine, but wings and tail relatively longer than Peregrine's, and wings more rounded than Peregrine's but less broad at base than Saker's. Upperparts rather dark grey-brown and barred; underparts variable from very pale, sometimes white, to heavily streaked. Moustachial stripe well-defined but narrower than Peregrine's. Adult has rusty-red nape and hind-neck unlike any of the other three species. Frequents cliffs, rocky slopes, ruins, stony plains deserts. Breeds southern Italy, Sicily, western Yugoslavia, Albania, Greece, Turkey, almost whole of north Africa and parts of Middle East. Largely sedentary, but occasionally recorded in other areas, for example Cyprus.

Fig. 55D. **Peregrine** *Falco peregrinus*. Large adult male shown. Smallest of the four, though size varies most (females larger than males), with more pointed wings and rather short tail. Upperparts of adults generally dark bluish-grey; underparts white to light reddish, densely barred blackish except for very pale chin and throat and lightly streaked upper breast. Upperparts of juveniles dark brown, underparts streaked brown. Black moustachial stripe (dark brown in juveniles) very broad and conspicuous. Almost ubiquitous, but requiring crags or cliffs, occasionally buildings or forests, for nesting. Breeds most of Europe (except Iceland), but has declined very much in last 20 years and now absent from several countries, e.g. Denmark, Netherlands, as well as from most of lowland Britain and Ireland. Also breeds Turkey, Caucasus, north coasts of Morocco and Algeria. Partly migratory, wandering birds (normally singly) from September to April anywhere in west Palearctic except Iceland, northern Fenno-Scandia and Russia and north Africa south of breeding range.

Gyrfalcon
Falco rusticolus
(plates 68, 69)

SILHOUETTE Wing-span 120–135 cm. Largest falcon, appreciably larger than most Peregrines (with which comparison most often made) and heavier in body, with longer, wider tail; long-winged, with broader and longer arms and broader hand with more rounded wing-tips.

FLIGHT A magnificent flier, often hunting close to ground, where it 'runs down' prey (plate 69), only occasionally stooping at it as Peregrine. Flight slower than Peregrine; wing-beats slower and shallower, appearing almost as if made by hands alone, but no real difference if speeding up. Soars and glides on wings straight out from body, held level (Fig. 53A) or slightly bowed, rarely with hands above level of arms.

IDENTIFICATION Distinguished from Peregrine by larger size, heavier build and different silhouette, though large individuals of northern race of Peregrine *F. p. calidus* closely approach small Gyrs. White phase is unmistakable; adults of grey and dark phase are pale or dark ash-grey, not so blue as Peregrine and, at least the grey phase, clearly paler above and usually with pale primary patch, absent in Peregrine. At close range, pale cross-bars or spots on upperparts and coverts of both grey and dark phase not seen in Peregrine. Spotting below is looser and 'rougher', with barring restricted to flanks and 'trousers'; Peregrine has denser and more regular barring. Adult Gyr has less contrasting head-pattern with barely noticeable moustachial streak, particularly in the rather pale headed grey-phase. Juvenile Gyr, especially of dark-phase, can be rather similar to juvenile Peregrine, but should be identified by combination of size, wing-shape, length of tail and less clear-cut moustachial streak which merges more with pale streaky patch on cheeks. However, young *F. p. calidus* (see Fig. 54) can be more like young Gyr, some being pale on forehead and crown with relatively pale brown upperparts and wing-coverts which have more pronounced, broader paler edging than other Peregrines in Europe; the moustachial streak in *calidus* is also narrow, but long and well contrasting with less streaky whitish patch on cheeks. Probably best character is underwing-pattern of Gyr which, as in adults, shows often more conspicuous contrast between densely dark-spotted underwing-coverts and pale fainter barred, more translucent primaries.

On average, Gyr has broader wings than Saker but flight-silhouette similar and size difference of no value in identifying lone birds. No Gyr has the Kestrel-like colours and contrasts above of adult and some young Sakers and adult Gyr, grey-phase has barred flanks and 'trousers', pale primary patch and whitish cross-bars above. Some young Sakers are grey-brown above without rusty colouration and are similar to young dark Gyr; young Saker has however larger, paler cheek-patch producing more distinct moustachial streak, and more pronounced tail barring. Perhaps best distinguished from Saker by less dense blackish-brown spotting on underwing-coverts and axillaries causing less contrast with pale flight-feathers.

Female Goshawk *Accipiter gentilis* may sometimes superficially resemble Gyr when wing-tip closed and thus looking rather pointed. However, long tail of Goshawk and different silhouette with comparatively short and broad wings, with rather curved trailing edge, diagnostic.

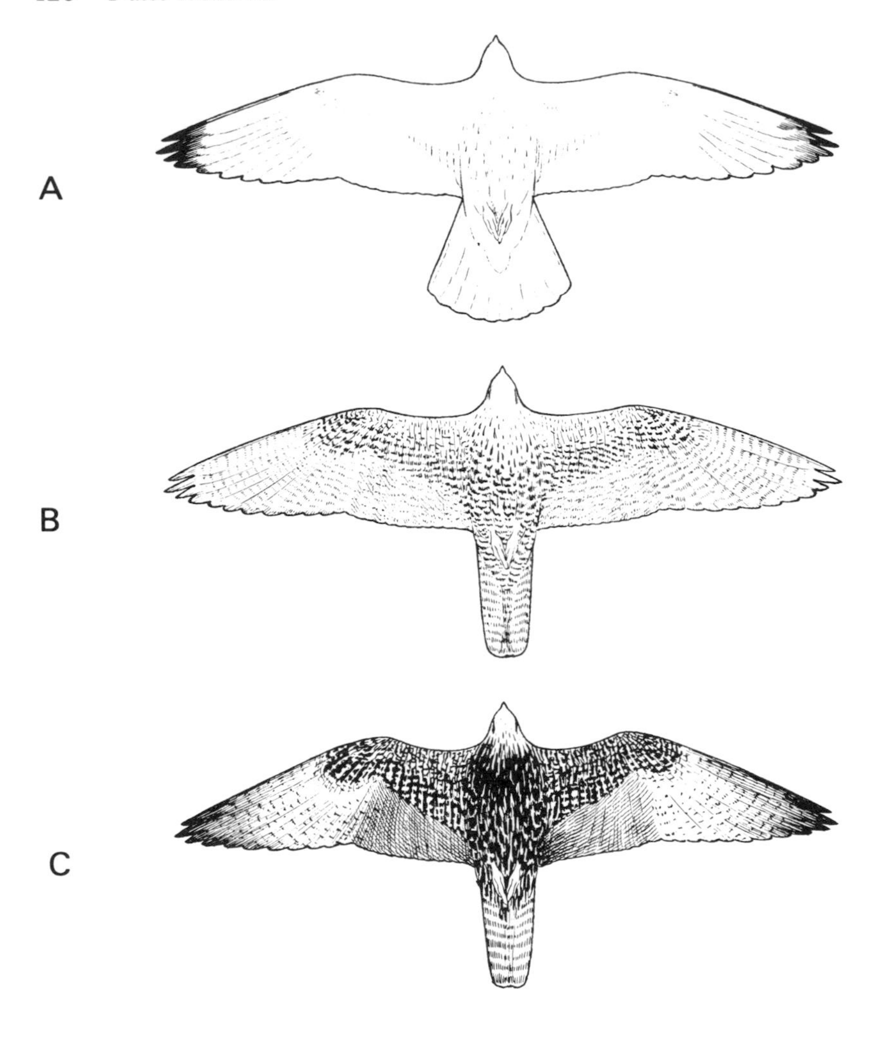

Fig. 56. **Gyrfalcon** *Falco rusticolus*. Typical individuals of adult white and grey phases and juvenile dark phase from below (see also caption to Fig. 55A). White phase adults (A) are unmistakably white all over with some faint dark streaks on sides of breast and belly and sparse dark spotting on flight-feathers and dark tips to primaries; apart from the last these markings are visible only at very close range. White phase juveniles (not shown) resemble adult but have darker wing-tips, more spotting on secondaries and tips to primary coverts, sparse streaks on breast and barred tail. Underparts of adult grey phase (B) are whitish, with unstreaked chin and throat, scattered brown drop-like spots on sides of neck and upperbreast, denser longitudinal spots on the lower breast and belly, and light bars on the flanks and tibial feathers. Whitish underwing-coverts are spotted or barred dark brown, usually less densely on the forewing-coverts. Flight-feathers are whitish, faintly barred and in field looking almost translucent and somewhat paler than coverts. Juvenile grey phase (not shown) has underwing-coverts more heavily spotted than adults, contrasting better with very pale flight-feathers especially primaries. Underparts of juvenile dark phase (C) are more densely spotted than in grey phase (adult or juvenile), particularly on underbody and wing-coverts, which appear dark and contrast with pale greyish faintly barred flight-feathers and tail. Adult dark phase (not shown) is paler than juvenile with less dense spotting on underbody and coverts.

Fig. 57. **Gyrfalcon** *Falco rusticolus* from above. White phase adult (A) is white with dark tips to primaries, lightly barred flight-feathers, faint dark tips to wing-coverts and mantle and faintly barred tail. Juvenile (not shown) has denser spotting on upperwing, particularly on coverts and mantle. Grey phase adult (B) has variably pale or dark slate-grey upperparts, with pale grey or whitish feather edgings giving a scaly appearance to wing-coverts, mantle, back and tail-coverts. Head is mainly white, narrowly streaked blackish; there is a pale eyebrow, diffuse dark eye-streak and usually a pale patch on the nape; cheeks are whitish, with faint moustachial streak. Primaries are grey-brown, with whitish cross-bars on the inner-webs which combine to produce pale patch on upperhand; secondaries are barred slate and whitish. Tail is distinctly barred dark brown and greyish-white. Juveniles of grey phase (not shown) are dark brown above, otherwise fairly similar to adult. Dark phase juvenile (C) is dark brown above with dark cheeks and ear-coverts almost merging into moustachial streak which is often well marked. Adult (not shown) is similar to juvenile but has dark brownish-slate upperparts, darker than in grey phase, with less pronounced pale feather edgings. Upperhand is slate-grey, much darker than in grey phase; cheeks and ear-coverts are also darker, with pale eye-brow and nape-patch less distinct; the moustachial stripe is diffuse but often darker and better marked towards throat.

Saker
Falco cherrug
(plates 70, 71)

Saker Falcon (saceroides type) attacking Little Egret

SILHOUETTE Wing-span 104–129 cm. Large, heavy falcon with wing-span almost as Buzzard *Buteo buteo*. Longer-winged than Peregrine, with somewhat longer arms, broader hands and more rounded wing-tips (seen when soaring). Wings often fairly broad at base, but narrower-winged examples not rare. Square-cut or slightly rounded tail fairly long, usually considerably longer than Peregrine's.

FLIGHT A fast, powerful flier, rarely hunting from great heights. 'Runs down' and stoops at prey in typical manner of other large falcons (but takes most prey on ground). Flies with slow, flat wing-beats; appears almost as if hands alone are moving. When hunting, wing-beats faster, stronger and deeper. Active flight often interspersed with short glides. Sometimes hovers like Kestrel *F. tinnunculus*. Soars on flat or slightly lowered wings held almost straight out from body, but on occasions soars or sails on wings slightly raised.

IDENTIFICATION Confusion most likely with Gyr and Lanner. Distinctions from Gyr already given under that species. Generally larger than Lanner, with heavier build and broader-based wings (compare plate 70 with plate 72). All white-headed Sakers easily recognisable but many have pale brownish-white crown with whitish nape to hind-neck, while Lanner has darkish brown crown with rusty-red on sides of rear edge and on nape to hind-neck. Very thin or almost no moustachial streak distinguish such Sakers, but otherwise little difference in this feature between two species in field. Kestrel-like, dull, rufous-brown upperwing-coverts contrasting with darker flight-feathers is not seen in adult Lanner, which has dark slate or ash-grey upperparts and wing-coverts; some Sakers even have yellowish-brown upperwing-coverts in strong contrast with dark flight-feathers and are unmistakable. Although spots and streaks on underbody average heavier and denser in adult Saker, difference often not obvious, since some Sakers have reduced streaking below; at close range the lack of transverse markings on flanks and 'trousers', seen in many adult Lanners, is only reliable character. Underwing pattern rather similar: adults of both species can show distinct dark band across rear coverts, but can be indistinct, or almost lacking, and both species can show all underwing-coverts evenly spotted; however, adult Saker has on average darker and more contrasting band on rear coverts and adult Lanner has on average more evenly-spotted wing-coverts. Flight-feathers below not conspicuously different and both species can show rather whitish translucent primaries, but on average Saker's flight-feathers are less conspicuously barred. At closest range adult Saker has a whitish supercilium running backwards from above eye (lacking in Lanner), less pronounced and clear-cut dark eye-streak and often a thinner moustachial streak. Brown back and wing-coverts in Saker have rusty feather edgings (in fresh plumage), slate-greyish in Lanner with rufous-grey cross-bars, most conspicuous on lower back to uppertail-coverts. In Saker uppertail medium to darkish brown with ochre-yellow cross-bars normally broken on central tail-feathers (which cover most of closed tail); in adult Lanner tail above is evenly and conspicuously barred ash grey and rufous-grey across all tail-feathers. Rare *saceroides*-type of Saker has more grey upperparts and wings (less contrast) with more or less rusty-yellow barring, which

(*continued p. 130*)

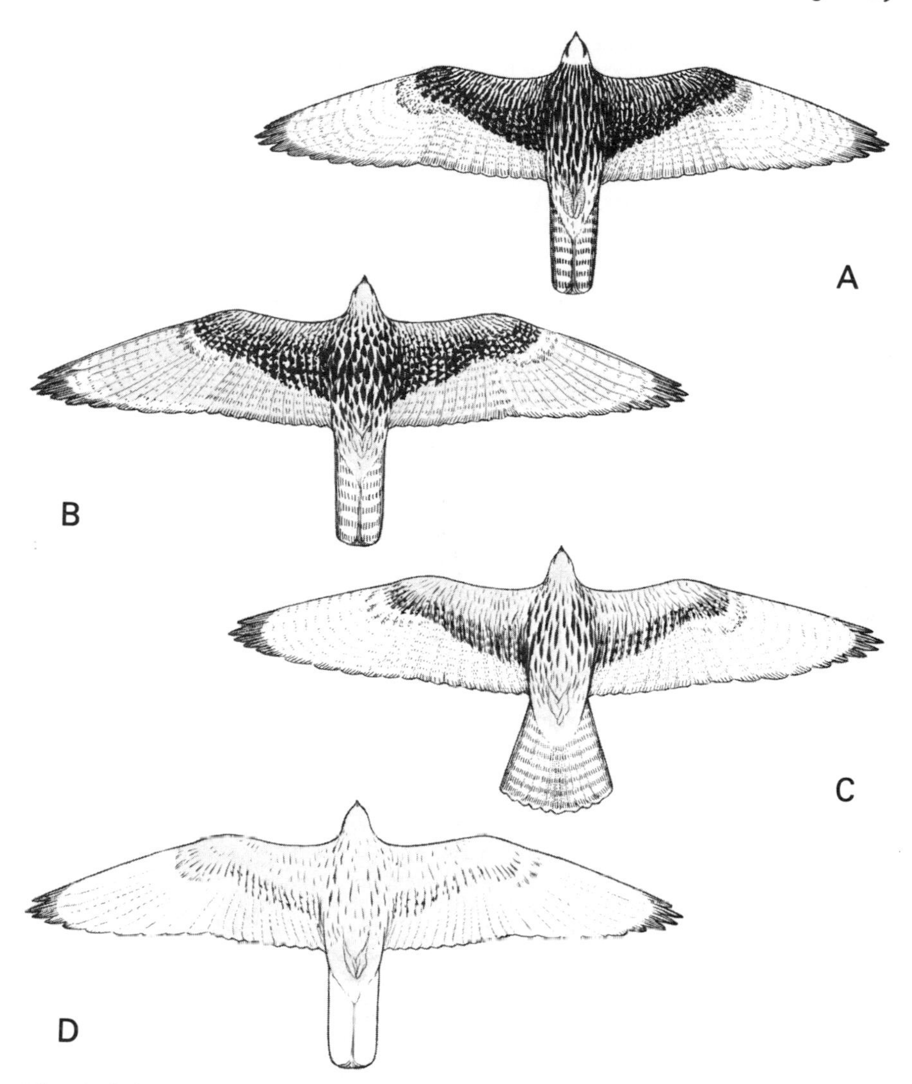

Fig. 58. **Saker** *Falco cherrug*. Juvenile (A) and three adults from below. East European birds described here belong to dark end of a cline within the nominate sub-species; by some authors they are separated as *F. c. cyanopus*. Underparts of adults are typically creamy-white, chin and throat unmarked, but remainder with dark drop- or spear-shaped streaking of markedly variable density (B and C). Some birds are very pale below (D), particularly *saceroides* morph, but also a few males of typical morph. Such birds may show faint barring on flight-feathers or tail, or, as depicted here, such faint barring as to be invisible in field. On typical birds, creamy and dark brown bars on undertail are of equal width. Flight-feathers, particularly primaries, are silvery-whitish with dark barring, darkest on secondaries; at close range flight-feathers, especially secondaries, are clearly barred, but at distance look very pale with darker wing-tip and trailing edge to primaries, and often appear translucent. Underwing-coverts and axillaries are creamy-white, streaked and spotted with variable amounts of dark brown, always much darker on greater secondary and lesser primary coverts, thus appearing at distance as broad, dark band along middle of wing. Some adults, notably of *saceroides* morph, have dark streaky/spotting on these areas much reduced (D), though it always appears as a darker band running lengthwise along underwing. Many juveniles (A) are separable from dark adults by blacker, more uniform underwing-coverts, and less drop-shaped streaking starting more abruptly below creamy throat, but a small percentage of young birds more resemble adult depicted in (B).

often also includes central tail-feathers; they must be distinguished primarily on head pattern (see Fig. 54). Some birds of this morph have very ill-defined paler barring on body, wings and tail and may look quite pale uniform dusty greyish-brown, or even greyish at a distance. Palest birds of this colour form always seem to have the whitest underparts and underwings. Juveniles are difficult to separate in field, and at distance often impossible; silhouette characters may be used with care only by experienced observers. Young Saker has paler top of head and slightly whiter hind-neck compared to Lanner but it often appears relatively dark-headed at a distance; patterns on sides of head slightly fainter, dark eye-streak less pronounced than in young Lanner. Moreover, Kestrel-like colours and contrasts above are frequently lacking, upperparts looking dark grey-brown with little darker flight-feathers; finally, juveniles of both species have uniform grey-brown central tail-feathers, though young Saker can show few small ochre-yellow spots.

In silhouette, Saker differs from Peregrine by generally larger size, longer tail, longer wings (longer arms) which are often more broad based and rounded at tips when soaring; active flight on stiffer, shallower wing beats. Problem of identification on plumage likely only with young Peregrines, which however have very dark upperhead heavier moustachial streak and darker, uniform upperparts. Large tundra race of Peregrine *F. p. calidus* may cause difficulties (see Peregrine and head-patterns Fig. 54). Juvenile Saker's underwing pattern with very dark coverts contrasting with almost translucent primaries is not found in Peregrine.

Fig. 59. **Saker** *Falco cherrug*. Adult from above (see also Fig. 54). Adult of east European form shown here, which is at dark end of cline (see caption to Fig. 58), has creamy-white crown finely streaked blackish, whitish nape-to-hind-neck with small diffuse blackish spot in centre, variably obvious whitish supercilium and dark eye-stripe extending to side of nape. Moustachial streak is also variable, either narrow but quite distinct stripe or very thin dark streak, hardly noticeable except at closest range. Large whitish cheek patch extends to eye. At distance details become obscure, whole head appearing whitish, or brownish-white crown with whitish hind-neck. Upperparts and wing-coverts are dark brown with rusty feather edges, in field looking reddish-brown and contrasting with blackish-brown primaries (secondaries are a little paler). Some adults appear almost yellowish-brown or pale grey-brown above, thus showing even greater contrast with flight-feathers. Uppertail is dark to rusty-brown with yellowish spots forming broken bars on outer feathers, partly or completely lacking on central ones; tail has pale rusty tip. Upperparts of juvenile (not shown) appear darker, showing less Kestrel-like contrast with dark flight-feathers than adult; crown is less pale than in adult as a result of coarser dark streaking, and moustachial streak is generally broader and longer, and pale markings on central tail-feathers are even more reduced. A rare phase occurs, so-called *saceroides*-type, in which colours of upperparts and wing-coverts are dusty greyish-brown, particularly on rump and uppertail-coverts, and there are variable pale rusty-yellowish cross-bars on back, scapulars and uppertail, less so on wing-coverts. Sometimes only few creamy spots arranged vaguely in lines across sides of uppertail. Some individuals of this morph look very uniformly greyish above at distance, and almost completely creamy-white below with only a dark area of streaking up centre of wing and faint dark streaking on body; barring on flight- and tail-feathers very faint indeed, difficult or impossible to see in field (see 58D also). In general appearance such birds often resemble adult Lanners, particularly races of Lanner inhabiting north Africa and Middle East.

Lanner
Falco biarmicus
(plate 72)

Lanner Falcon chasing Rock Dove

SILHOUETTE Wing-span 95–115 cm. Medium-sized falcon, size between Saker and Peregrine. Compared with Peregrine has longer and slightly narrower wings, hand a little broader and wing-tips less pointed, tail generally slightly longer.

FLIGHT Graceful, as Peregrine, though less bold and powerful. Takes much of prey in air, and capable of tremendous stoops. Flies with slow, flat wing-beats, accelerating when hunting to faster and deeper wing-beats. Shows great manoeuvreability in leisured flight at breeding cliffs, performing fast gliding on angled wings. Soars on flat or slightly lowered wings like other falcons.

IDENTIFICATION Adult of European race *F. b. feldeggii* usually distinguishable from adult Saker by lighter build, generally narrower wings and darker, rusty-red nape and hind-neck. Upperparts often darker than Saker's, more or less slaty, without Kestrel-like colour pattern; at close quarters pale bars on shoulders, back, rump and uppertail-coverts, as well as prominent narrow tail-barring, are good characters. Underparts generally less heavily marked than Saker's, some males even entirely white below. Often shows bars on flanks and tibial parts, a character not found in Saker. Detailed distinctions between head and underwing patterns of Lanner and Saker, and criteria for separating juveniles, already given on pages 128 and 129.

Lanner differs from Peregrine in being longer-tailed and slightly longer-winged and in having less pointed wing-tip and slightly broader hand when soaring. Adults differ in colour of upperhead, width of moustachial stripe, extent of white cheek patch, and markings on underparts. Juveniles more difficult to distinguish; both dark on upperhead, but Lanner's moustachial stripe generally narrower and better defined, and difference in white cheek patch same as in adults. Both show longitudinal streaking on underparts, but Lanner's generally heavier, contrasting more with unmarked throat. Dark underwing-coverts of young Lanner show far more contrast with flight-feathers than do those of young Peregrine. See Peregrine for separation from juvenile and adult *F. p. calidus*.

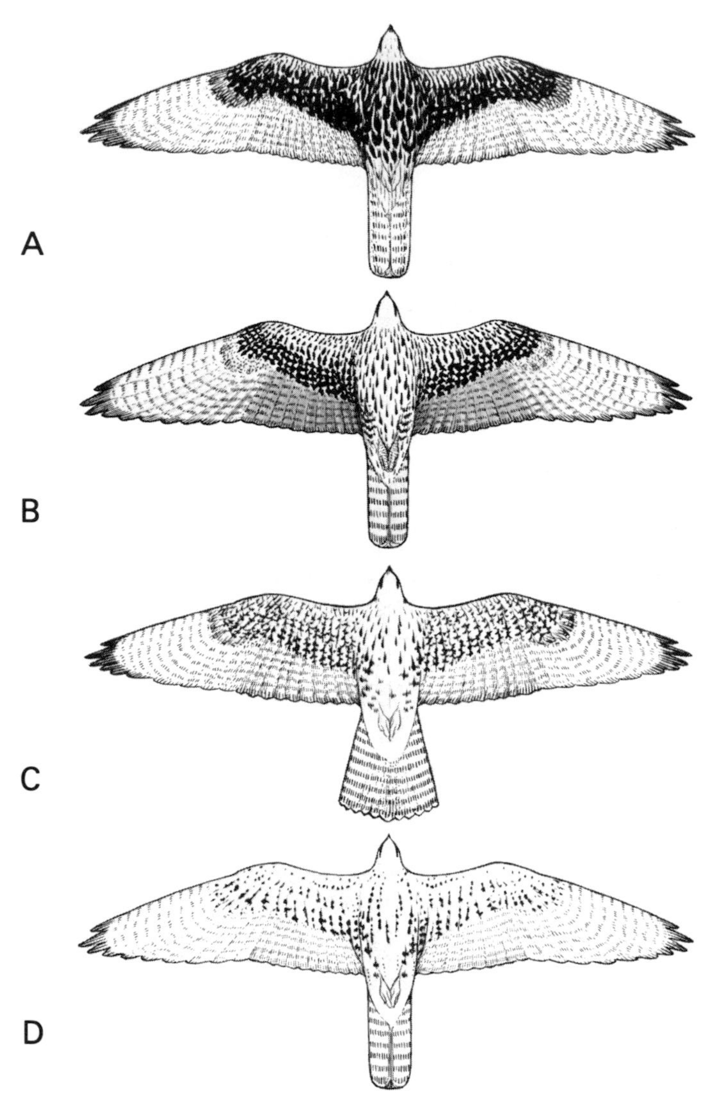

Fig. 60. **Lanner** *Falco biarmicus*. Juvenile (A) and three adults from below. Underparts of adults (B, C and D) are creamy or whitish, with small drop-like or heart-shaped black spots on sides of breast, becoming larger on belly and often appearing as bars or semi-barring on flanks and 'trousers'. Some males, however, appear all-white below without dark spots. In others, spotting covers breast and belly, but not thighs which then appear pale creamy and quite distinct from rest of underparts, particularly when perched. Underwing-coverts are creamy with dark brown spots, in some fairly evenly spotted—thinly or densely—but in others more heavily marked to form dark band on coverts, while body more thickly streaked (B). Flight-feathers are greyish-white below with more or less pronounced dark grey bars, in certain lights primaries appearing translucent, except for darker wing-tip, while at distance flight-feathers (and with pale birds whole underwing) can look very pale. At close range tail evenly and distinctly barred dark and pale with whitish tip, but at distance looking pale.

Underparts of juvenile (A) are whitish with brownish hue, heavily marked with broad, longitudinal streaks, heaviest on flanks and can appear all dark in contrast with whitish throat. Underwing generally shows more contrast than in adult, coverts being heavily dark-spotted, often as blacker band along rear coverts and contrasting with greyish-barred flight-feathers. Undertail as adult, but barring less clear-cut.

Fig. 61. **Lanner Falcon** *Falco biarmicus*. Adult from above (see also Fig. 54). Adult of European sub-species *F. b. feldeggii* has rusty-red crown to hind-neck, densely streaked black on crown and creamy forehead. There is blackish-brown eye-streak from eye to side of nape, and long white cheek patch extending up to eye. Moustache is well-defined in narrow, long, black streak. Upperparts and wing-coverts slate or ash-grey, feathers edged greyish-red, creating barred appearance at close range; barring paler and more pronounced on scapulars, rump and uppertail-coverts. Colour of upperparts and wing-coverts contrasts with dark grey-brown flight-feathers, darkest on primaries. Uppertail is ash-grey with regular, unbroken, reddish-grey or pale grey bars and whitish tip. Juvenile (not shown) differs by having considerably darker (almost blackish-brown) upperhead, feathers edged rufous, sometimes only on nape and hind-neck. Upperparts and upperwing-coverts dark slate-brown with pale feather edgings, narrower than in adult and without transverse barring, thus being generally dark and not contrasting with flight-feathers; pale cheek patch is finely streaked black. Whilst outer tail-feathers are barred, central ones are not, being plain grey-brown; thus when perched no tail-barring is visible.

Peregrine
Falco peregrinus
(plate 73)

Peregrine hunting Wood Pigeons

SILHOUETTE Wing-span 80–117 cm. Medium-sized, compact falcon with relatively short, broad-based arm, but narrow hand and more pointed wing-tip than other large falcons; wings often appear triangular-shaped. Square-cut or slightly tapering tail is of medium length, but slightly longer in juvenile generally. Female markedly larger than male and some of large tundra race *F. p. calidus,* which move south in winter, approach male Gyr in wing-span.

FLIGHT A very powerful, swift and agile flier. Active flight with quiet, but solid and stiff, not very deep, wing-beats, now and then interrupted by short glides. When hunting, wing-beats faster, stronger and deeper. Often hunts from greater heights than other falcons, 'runs down' prey and stoops at it with great speed on partly or nearly fully closed wings. May be seen gliding to and from cliffs on angled wings, or maintaining position in updraught on slightly lowered wings. Rarely hovers like Kestrel. Soars with front edge of wings slightly angled, rear edge more straight out from body; wings held level or slightly lowered when soaring and gliding.

IDENTIFICATION Adult separable from adults of other three species by dark bluish- or slate-grey upperparts, white to light reddish underparts densely barred blackish, and white cheeks, contrasting strongly with very conspicuous and well-defined black moustachial 'lobe' and black top of head. Further differs from Gyr and Saker by generally shorter tail, shorter and more triangular-shaped and more pointed wings. Some pale juveniles of *F. p. calidus* (see Fig. 54) much resemble some adult Sakers in having narrow but long moustachial stripe, very large white patch on cheeks-ear-coverts, narrow, dark eye-streak, whitish forehead, supercilium and nape, just with crown streaked blackish; furthermore, they can be paler medium grey-brown above with contrasting darker flight-feathers. Darkish brown hind-neck in *F. p. calidus* (whitish in adult Saker, which is sometimes white-headed), lack of rufous colouration above (though sometimes less apparent in some bleached and all *saceroides* morphs of Sakers) and less conspicuous pale barring in spread uppertail, together with lack of contrasting underwing pattern, seen in most Sakers, distinguish *F. p. calidus.* Distinguished from young dark-headed Sakers by normally much less streaked underparts, lack of contrasting underwing-pattern and thinner pale tail-bars above. Silhouette characters are also useful (which see). For distinction of *F. p. calidus* from young Gyr, see that species.

Flight silhouette may recall Lanner, but Peregrine is more compact, less elongated. Soaring Peregrine shows more triangular-shaped wings, narrower hand and more pointed wing-tip. Adult has blacker top of head, darker forehead and blackish hind-neck (creamy forehead and rusty-red hind-neck in adult Lanner), much broader moustachial streak, thus smaller white cheek-patch, less pronounced barring on

(*continued p. 136*)

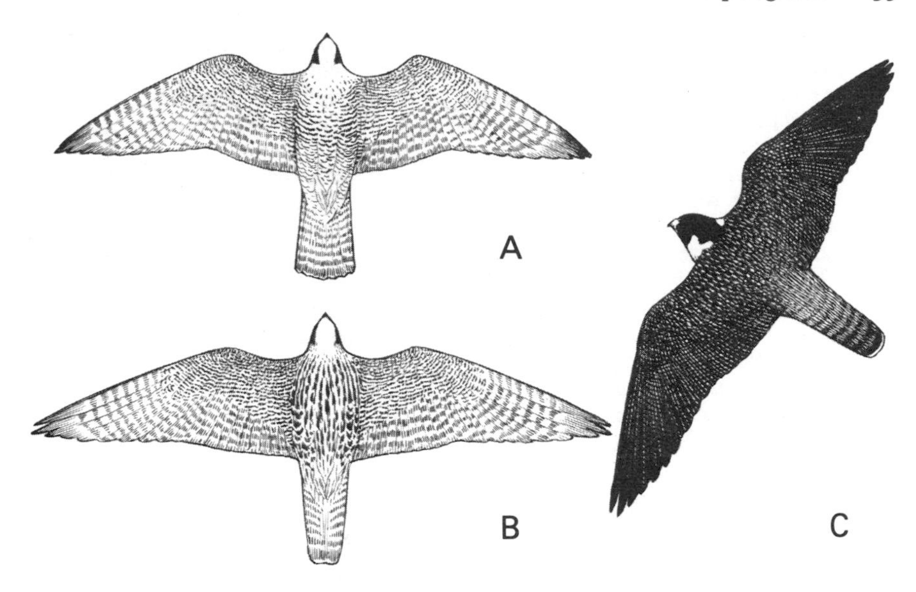

Fig. 62. **Peregrine** *Falco peregrinus*. Adult and juvenile from below and adult from above. Three subspecies breed in Europe: widespread nominate race (illustrated here), paler tundra race *F. p. calidus*, which moves south in winter, and smallest *F. p. brookei* from the Mediterranean region. Adult of nominate race (A and C) has white underparts (often with faint warm buffish cast on breast), densely barred blackish on lower breast to undertail and underwing-coverts; females are often more heavily barred. Flight- and tail-feathers below greyish-white with dark grey barring; tail tipped whitish. At distance underwings look greyish, though primaries sometimes look paler and semi-translucent with darker wing-tip. Upperparts slate-black on upperhead, nape and lesser wing-coverts, bluish- or slate-grey on remainder of upperparts (paler on rump and uppertail-coverts) and blackish on flight-feathers. At distance upperparts look very dark except lower back and rump, but close views reveal dark barring, especially on tail which is fairly strongly barred slate-grey and blackish-brown with whitish tip. Eye region and ear-coverts blackish, as is broad, moustachial 'lobe' which contrasts with white cheek.

Juvenile of nominate race below (B) has yellowish-white or pale rusty-buff body with dark brown streaks except on chin and throat; underwing-coverts coarsely barred, spotted and streaked, while flight-feathers barred much like adult's. At distance underwings appear brownish with slightly paler primaries. Above it is blackish-brown, slightly paler on rump to uppertail-coverts and at close ranges shows narrow, rusty-yellow feather-edgings, forming bars on uppertail-coverts; dark brown tail has narrow, pale barring; there is small buffish patch on nape, cheeks are yellowish and often finely streaked and blackish-brown moustachial streak is narrower than in adult.

Adult of smaller *F. p. brookei* (not shown) resembles nominate race but is more rufous-pink below, generally with more dense barring; above it differs in often having a broader moustachial 'lobe', frequently covering most of cheeks, and usually dull rufous patch on nape which occasionally extends onto crown.

Juvenile *brookei* is darker below, more heavily streaked; above, pale nape-spot is more rufous-brown than nominate juvenile.

Adult of tundra race *F. p. calidus* is whiter on underparts, often with considerably fainter barring, particularly on underwing-coverts; in some barring restricted to flanks and 'trousers'. Above is generally paler grey or blue-grey, especially on head, moustachial streak usually narrow and long and the white cheek-patch behind often larger (Fig. 54).

Juvenile *calidus* is generally less heavily streaked below and particularly pale birds white below with very thin, scattered, brownish streaking; above is paler brown than nominate race juvenile, often in contrast with dark flight-feathers; much overlap with features of nominate juvenile but, typically, has paler edgings to feathers of upperparts, paler forehead, more pronounced, whiter nape-spot, whitish supercilium, narrow moustachial streak and larger area of white on cheeks; markings on head can resemble those of other large falcons (see Fig. 54).

upperparts and denser barring on belly. However, a few adult *calidus* (see Fig. 54) have paler top of head, narrower moustache and larger white patch on cheeks and they can have barring below restricted to flanks and 'trousers', thinly barred underwing-coverts and paler blue-grey upperparts; such pale adult Peregrines lack adult Lanners' rusty-red hind-neck and well-defined black eye-streak and do not show dark band on rear underwing-coverts of some adult Lanners. Juvenile *calidus*, particularly pale birds, harder to separate; distinctions already given under Lanner (and above for separation from Saker); they always lack rusty-red hind-neck, have thinner streaking below and paler underwing-coverts without contrasting dark band on rear coverts of young Lanner.

Confusion possible with Eleonora's Falcon *F. eleonorae*, but that species much more elegant with slender build and distinctly longer tail, and also considerably darker underparts (even pale phase). Confusion with Hobby *F. subbuteo* possible only when size difference not apparent, but slimmer Hobby has narrower, more scythe-like wings which are darker below.

6 The Small Falcons and Black-winged Kite

Five of the seven species dealt with in this group—Kestrel *Falco tinnunculus*, Lesser Kestrel *F. naumanni*, Hobby *F. subbuteo*, Red-footed Falcon *F. vespertinus*, and Merlin *F. columbarius*—are widespread and fairly common in their respective breeding areas. Of the remaining two, Eleonora's Falcon *F. eleonorae* is scarce and mainly restricted to islands in the Mediterranean region, and Black-winged Kite *Elanus caeruleus*, though widespread in Africa and Asia, is found in Europe only in Portugal and parts of Southern Spain.

The females of the two kestrels can be easily confused though there should be no difficulty in the identification of the males. The habitually hovering Kestrel is one of the most widespread and common birds of prey in Europe.

The Merlin is usually readily distinguished by its small size alone, but the Hobby and Red-footed Falcon may be confused if plumage characters are not seen. The largest falcon in the group, Eleonora's, has two distinct colour phases—one light and Hobby-like, the other dark, superficially resembling the male Red-footed. As with other falcons, the underwing-pattern is important for field identification, and the degree of contrast on the upper-wing often proves very helpful. Five of the species—Lesser Kestrel, Hobby, Red-footed Falcon, Merlin and Eleonora's Falcon specialise in catching small birds or insects in flight, often consuming their prey, held in the feet, while on the wing. Eleonora's Falcon is highly adapted to a seasonal food supply by breeding in late summer and feeding its nestlings on migrant birds.

The Kestrel and Black-winged Kite hunt mainly by pouncing on rodents or insects on the ground, having located them from a vantage post or during their hovering hunting flight.

For convenience this group has been dealt with in two parts:

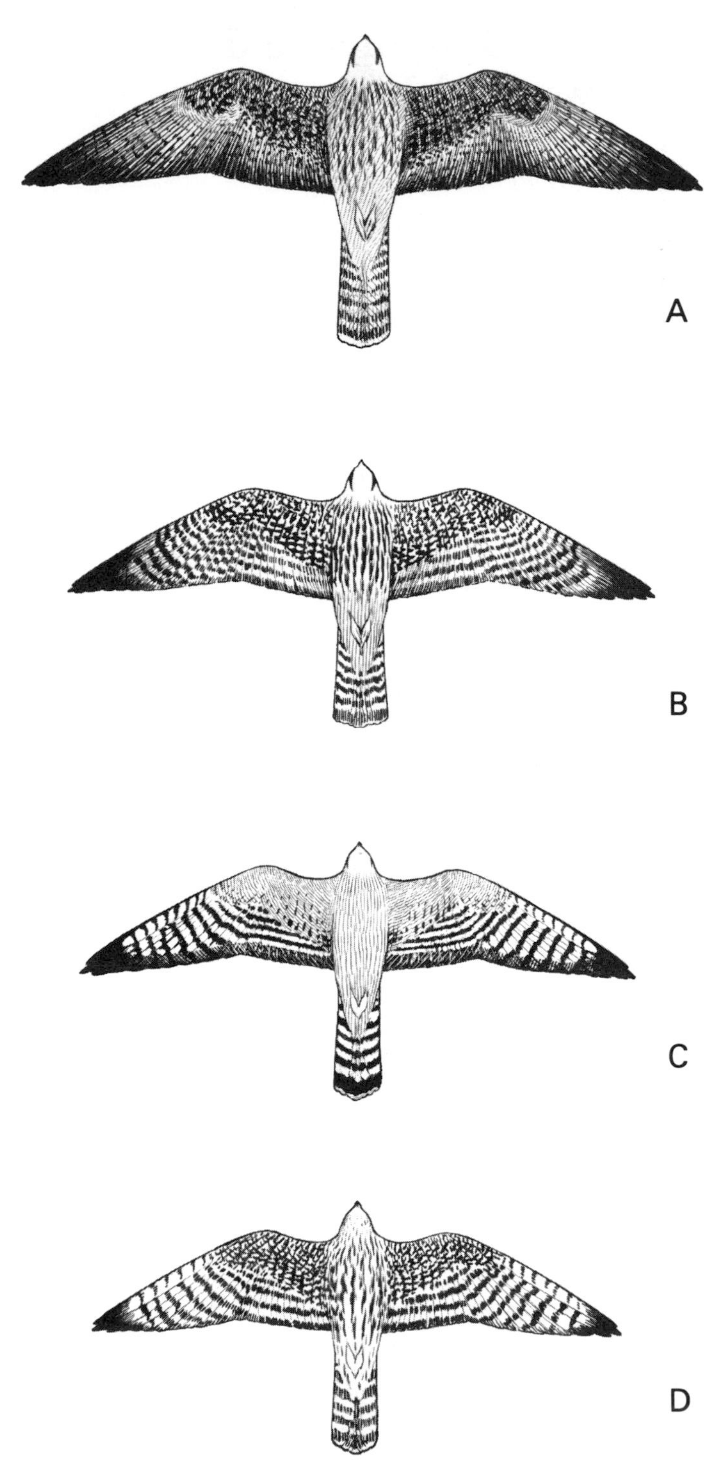

Fig. 63. Typical undersides of juvenile Eleonora's Falcon, Hobby, female Red-footed Falcon and female Merlin.

Fig. 63A. **Eleonora's Falcon** *Falco eleonorae*. Juvenile bird is shown. Medium-large falcon, in shape like large Hobby but with relatively longer wings and tail; foraging flight with slow, relaxed wing-beats, hunting flight dashing and agile. Underwing-coverts dark (spotted in juvenile), flight-feathers pale at base and darkening towards rear edge. At a distance, upperparts of light phase recall dark female Kestrel *F. tinnunculus*, particularly in strong sunlight. Dark phase sooty-grey, more uniform than adult male Red-footed. Juvenile browner above, with slightly shorter tail and less pointed wings than adult. Breeds colonially on sea-cliffs and rocky islets, July–September, eastern Canaries, off western Morocco, and in Mediterranean region, e.g. Balearic islands, Corsica (?), Sardinia, Elba, islands off Sicily and in Greek waters (including Crete and Rhodes), Cyprus, north-east coasts of Tunisia and Algeria; migrants occur in western and southern Turkey and may breed there. Migratory, April–May and October, though a few seem resident in the Mediterranean.

Fig. 63B. **Hobby** *Falco subbuteo*. Adult bird is shown. Small to medium-sized falcon with long, narrow, very pointed wings and short tail; a swift and elegant flier, catching prey in the air. Upperparts uniform dark slate; well-defined dark moustachial stripe. Underparts heavily streaked, tibial feathers and undertail-coverts unstreaked red in adults; underwing whitish, densely spotted on coverts and barred on flight-feathers, looking greyish in field. Juvenile dark grey-brown above, heavily streaked below. Breeds singly in wooded areas from south Finland, south Sweden, Denmark (rare) and southern England throughout continental Europe to Mediterranean and some of its islands, e.g. Corsica, Sardinia (?), Sicily, Crete (?) and Cyprus (very rare), also Turkey, Morocco, Algeria and Tunisia; migratory, April–May and late August–mid October.

Fig. 63C. **Red-footed Falcon** *Falco vespertinus*. Adult female is depicted. Rather small, slender and finely built, much like Hobby but with longer tail. Adult male dark sooty-grey, flight-feathers paler above, thighs and undertail-coverts red; underwing-coverts very dark and paler slate-grey flight-feathers. Female has unstreaked rufous-yellow head, body and underwing-coverts (head often bleached paler), pale greyish dark-barred back and upperwing-coverts, and darker flight-feathers. Juvenile similar, but with darker upperhead, densely streaked underparts, well-developed moustachial stripe and dark trailing edge to pale underwing. Breeds colonially in rookeries, Hungary, Romania and west Russia, also regular easternmost Austria and southernmost Czechoslovakia, irregular and scarce Poland, Germany, Baltic States, Yugoslavia and Bulgaria, rare Finland and Sweden; migratory, April–May and August–mid October, mainly south-east Europe, Turkey and eastern Mediterranean islands, slightly farther west in spring.

Fig. 63D. **Merlin** *Falco columbarius*. Adult female is shown. Smallest European falcon, with relatively shorter wings than other species, and longer tail than Hobby; flight bold and dashing, with rapid, stiff wing-beats. Male (almost as small as Mistle Thrush *Turdus viscivorus*) slate bluish-grey above with darker flight-feathers and broad blackish bar at tip of tail. Nape and underparts pale reddish-brown to buff, streaked black; no conspicuous moustachial stripe. Female larger, darker brown above; whitish streaked brownish below, uppertail with narrow dark bars and a broad one at tip; juvenile similar. Subspecies *aesalon* breeds open forest near wetlands, Ireland and north and west Britain (where disperses more widely in winter), and Norway, most of Sweden, Finland, Baltic States and north Russia (continental birds migratory, September–October and March–May, to most of Europe including Britain); *subaesalon* mainly resident Iceland, but some migrate into and through Britain and Ireland.

Eleonora's Falcon
Falco eleonorae
(plates 74, 75)

SILHOUETTE Wing-span 90 cm. Largest falcon of the six in this group, size between Peregrine *F. peregrinus* and Hobby, wing-span similar to male Peregrine. Finely built with narrow oblong body and rather long tail, longer and more rounded than in the other species. Wings relatively rather narrower and longer than those of Hobby, which it otherwise resembles most nearly in outline. Juveniles have less pointed wings and sometimes slightly shorter tail. Dark phase frequently appears slimmer than light phase.

FLIGHT An extremely good flier, very agile and swift, capable of tremendous speed and stoops. Wing-beats fast and regular when hunting, but also has a very characteristic slow foraging flight with peculiarly relaxed, flat, soft wing-beats, whole wing action slower than one would expect of a falcon of this size. The bird appears always to have more than enough time for comfortable flight. Hunts insects actively like Hobby, but wing action slower. Soars on flat or slightly lowered wings straight out from body, tail partly spread or closed. Spends much time gliding slowly on almost rigid wings above breeding cliffs.

IDENTIFICATION Confusion possible mainly with Hobby and young Peregrine, the dark phase—which accounts for about a quarter of all birds throughout the range—with male Red-footed. Differs in unique flight with relaxed, soft, slow wing-beats (but dashing when hunting), narrower and longer wings, and longer and usually more rounded tail. Dark phase is distinguished from male Red-footed by palish bases to flight-feathers below, darkening towards trailing edge (uniform in Red-footed), by lack of red 'trousers', vent and undertail-coverts; and from above by rather uniform upperwing surfaces (in Red-footed flight-feathers silvery-grey, contrasting with darker coverts). Compared with Hobby light phase has much contrast between very dark underwing-coverts and pale barred flight-feathers below (rather uniform underwing in Hobby), and by darker rusty-brown or rufous-buff underparts generally; dark individuals of pale phase also have darker throat and cheeks and are considerably darker on underparts than any Hobby. Adults distinguished from Peregrine by contrasting underwing-pattern which is much more uniform in Peregrine and distinctly barred at close range. Juvenile Eleonora's is distinguished from juvenile Hobby and Peregrine, apart from silhouette-characters, by much more pronounced and broader dark trailing edge to underwings, and by generally more contrast between paler flight-feathers and densely spotted brownish-looking underwing-coverts (less contrasting in both young Hobby and Peregrine); also, streaking below is narrower, particularly compared to Peregrine.

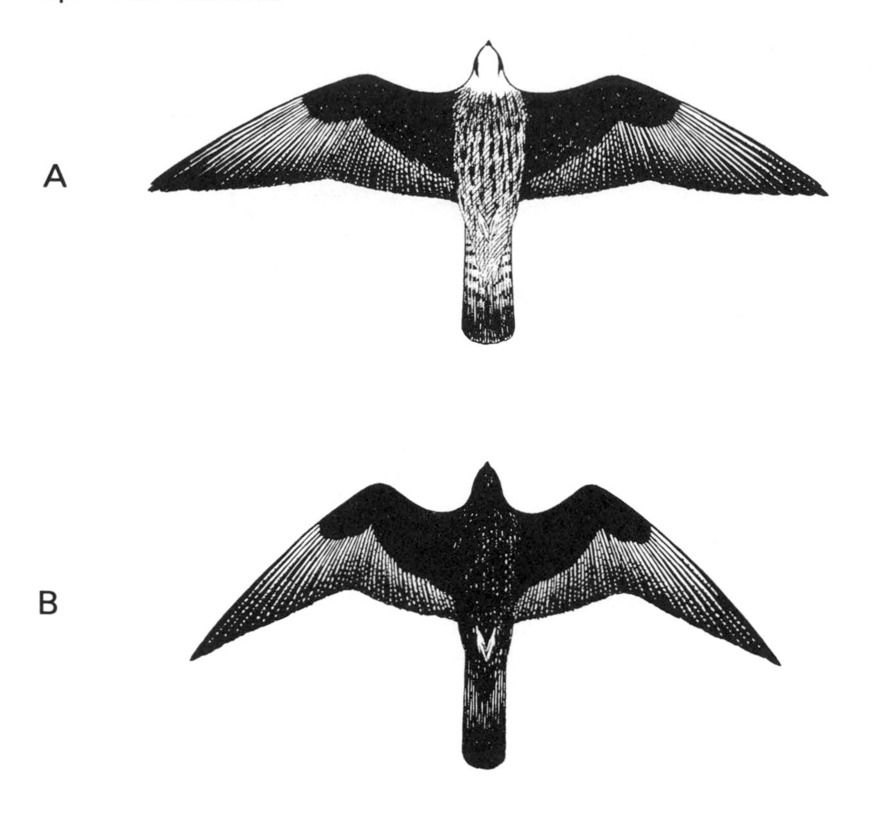

Fig. 64. **Eleonora's Falcon** *Falco eleonorae*. Light and dark phase adult from below. Light phase adults (A) have the chin and throat and a patch behind the dark moustachial stripe unstreaked whitish-buff, the remainder of the underparts rufous-buff to cream-coloured with a pink hue, densely streaked black on the breast, less so on the belly. The undertail-coverts look paler than the body, though sometimes they are uniform reddish-brown, finely streaked. The undertail is greyish-white with faint brownish bars. Dark phase adults (B) have the whole head and underparts dark slate-grey, the undertail somewhat paler, and at a distance often appear almost black. The underwing-coverts of both phases are very dark brown (more sooty in dark phase), flight-feathers either considerably paler greyish (perhaps larger contrast in dark phase), or more commonly, medium-grey at their bases, merging into a darker hind-wing and producing a diffuse pale area, sometimes as a line to basal secondaries, paler greyish-white as a diffuse patch on the primaries with darker wing-tip. Birds in intermediate plumage, resembling dark individuals of the light phase, are perhaps immatures in their first year: they often show rather dirty underparts, paler throat and cheeks, densely spotted underwing-coverts and rather pale barred flight-feathers that darken towards the rear edge. First-autumn juveniles of both phases (see A) are more brownish than light phase adults. Their underparts are less clear-cut, more broadly streaked on the breast but more finely on the belly; their underwing-coverts are pale with numerous dark spots, looking dark brown in the field; their flight-feathers are considerably paler beneath than those of adults, with many faint dark bars and without the effect of a contrasting pale line on the underwing, but with a characteristic wide blackish band along the trailing edge, and blackish wing-tips. The undertail is more clearly barred than in light phase adults, and has a whitish tip. The legs and feet are pale yellow at all ages.

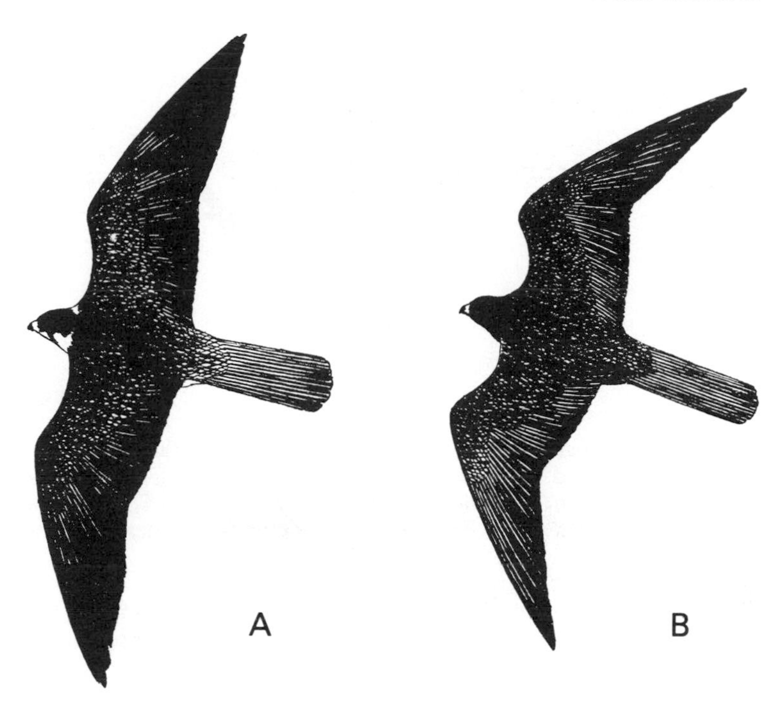

Fig. 65. **Eleonora's Falcon** *Falco eleonorae*. Light and dark phase adult from above. Light phase adults (A) are rather dark, uniform brownish-black or bluish-black on the upperparts, very little paler on the tail (and sometimes on the back or wing-coverts, or both), and slightly paler, but still very dark ashy-brown, on the upperhead. The dark brown moustachial stripe is usually a little wider than that of the Hobby. Dark phase adults (B) are dark slate-grey on the whole upperparts, practically without shade differences, and often appear almost black at a distance. First-autumn juveniles of both phases (not shown) have the upperwing-coverts and the feathers of the back edged pale buff. They are generally dirtier and browner above, and have more diffuse, less contrasting, head markings, than light-phase adults. At all ages, the cere is pale yellow, and the bill dark horn with a blackish tip.

Hobby
Falco subbuteo
(plates 76, 77)

Hobby chasing Swifts

SILHOUETTE Wing-span 69–84 cm. Medium-sized, slender falcon with narrow, scythe-like, sharply pointed wings, shape not unlike Swift *Apus apus*. Tail relatively short, square-cut and often with middle rectrices a little elongated, thus appearing slightly wedge-shaped when closed. Female larger and slightly broader-winged at base than male.

FLIGHT A swift, agile and elegant flier, catching prey in the air. When hunting, wing-beats fast, stiff and regular, interrupted by short fast glides and ending with lightning stoop on partly or almost completely closed wings. Foraging or migratory flight slower, with flatter, stiff wing-beats. When catching insects, movements quiet and graceful—mixture of fast hunting flight, gliding, and soaring on very pointed, rather Swift-like wings, directed slightly backwards; tail closed. Soaring on thermals typically on fully outstretched wings, straight or pressed a little forward (but angled at carpal joint) and flat or slightly lowered; tail more or less spread. Hovers only very occasionally.

IDENTIFICATION Very similar in flight silhouette to Red-footed, though a little larger and heavier built, and tail generally shorter (only last point may be diagnostic). Hobby's active flight lacks Kestrel-like loose wing-action, frequently seen in foraging Red-footed, and it very seldom hovers which Red-footed does often. Conspicuous head-pattern, heavily streaked underparts and unbarred uppertail distinguish Hobby from adult female Red-footed (Fig. 67C) and from first summer male (Fig. 68C) (which is either rufous with scattered streaks or later, unstreaked pale ash-grey on underparts). Compared with juvenile Red-footed (Figs. 67D, 68D), Hobby is much darker and plainer above (without contrast between coverts and flight-feathers of young Red-footed), with larger moustachial stripe, darker forehead and darker unbarred uppertail at close range, less yellowish on body and underwing-coverts, and uniformly greyish on underwing (spotted and barred at close range) without conspicuous dark band to trailing edge. Adult Hobby further separated from all these by well-defined red thighs and undertail-coverts.

Distinguished from Eleonora's by clearly shorter, usually more square-cut tail and relatively shorter wings, faster and stiffer wing-beats, and quite different movements when foraging; for plumage differences see page 141. Told from Peregrine by narrower wings, smaller size, finer build and more graceful flight (less heavy and bold); confusion on plumage likely to arise only with young Peregrine, but moustachial stripe narrower and uppertail unbarred. Differs from the two kestrels in wing action and flight, colour and lack of contrasts above, darker underwing, narrower and more pointed wings and shorter, squarer tail. Distinguished from Merlin in silhouette by larger size, relatively shorter tail but clearly longer wings which are more scythe-like, and by generally slower wing-beats and mode of flight.

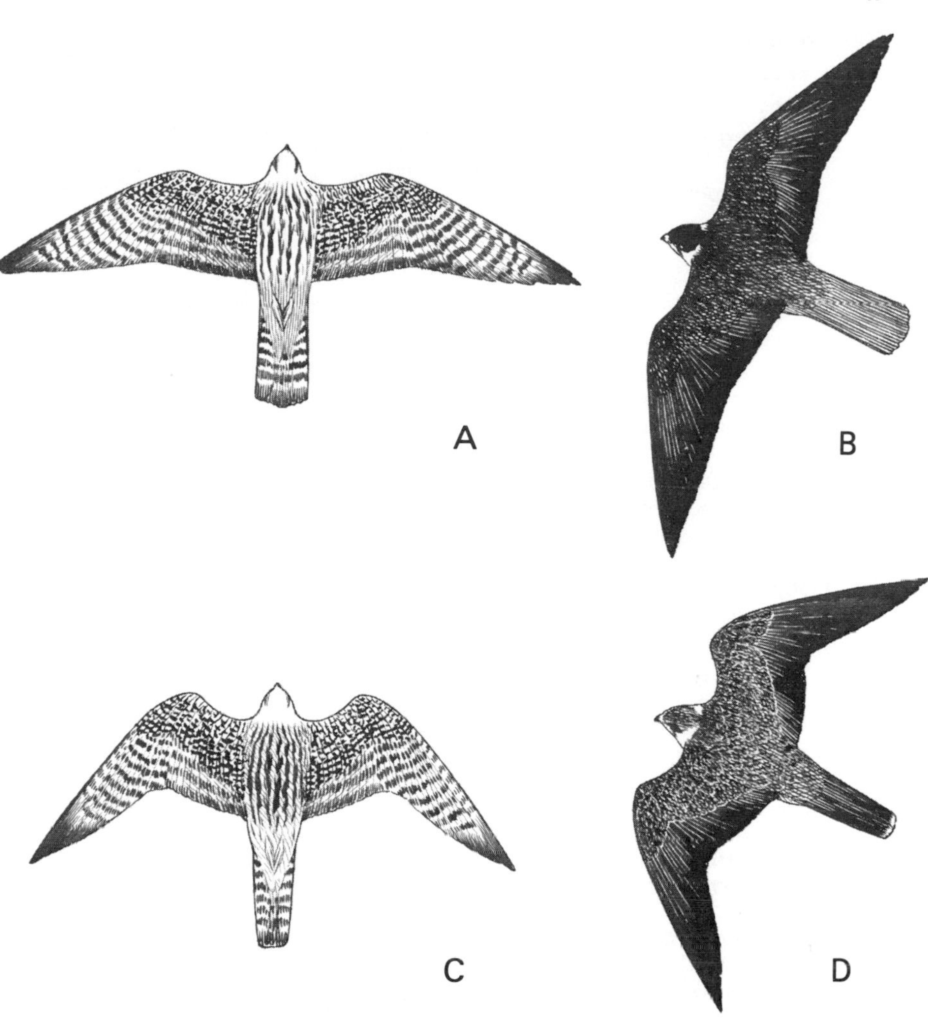

Fig. 66. **Hobby** *Falco subbuteo*. Adult and juvenile from below and above. Adults (A and B) have whitish to pale rusty underparts with broad blackish streaking, heaviest on the breast, unstreaked red tibial feathers and undertail-coverts, and a greyish-white undertail with indistinct dark bars. The underwing-coverts are creamy-white to buff with dense dark spots, the flight-feathers greyish-white with narrow dark bars and darker wing-tips; at a distance the underwing appears evenly greyish, paler or darker depending on light conditions. The entire upperparts are almost uniformly dark slate bluish-grey, palest on the upperhead, lower back, rump and unbarred tail, darkest on the mantle; females show a brownish tinge. The flight-feathers above are blackish-brown, a trace darker but not contrasting with the coverts. A long, narrow, dark moustachial stripe stands out against a whitish to yellow-buff patch behind. The legs and cere are yellow, the bill greyish, darkest at the tip. Juveniles (C and D) resemble adults very closely except for buffish tibial feathers and undertail-coverts. The underparts are more heavily streaked, though this is not a useful field mark. The upperparts are more brownish-black, the feathers of the upperwing-coverts, shoulders, rump and uppertail-coverts edged pale buff, being noticeable on tops of greater coverts as thin pale line at closer ranges; tips of flight-feathers also pale (showing as narrow pale line on trailing-edge) and tail more conspicuously tipped creamy than in adult.

Red-footed Falcon
Falco vespertinus
(plates 78, 79)

Three Red-footed Falcons (left) with Hobby

SILHOUETTE Wing-span 58–70 cm. Rather small and finely built, a little smaller than Hobby which it resembles in outline, though the narrow pointed wings are slightly broader at the base. Square-cut, slightly rounded or tapering tail a little longer than Hobby's, sometimes approaching that of Kestrel in length.

FLIGHT Graceful and agile, with fast, stiff, regular wing-beats when hunting, very like Hobby but with wing action faster. Movements graceful when catching insects, combining soaring, short glides and active flight, slow foraging flight often on rather Kestrel-like 'loose' wing-beats. Glides on scythe-like wings with closed tail, resembling Hobby. Soars on slightly lowered wings held straight out from body or pressed slightly forward, with tail more or less spread. Often hovers over fields on rapid, flat wing-beats, like Kestrel but less persistently.

IDENTIFICATION Easy to confuse with Hobby, from which differs in finer build, faster wing-beats and longer tail, but such characters of dubious value due to size overlap and variation in length of tail. Adult male (Figs. 67A, 68A) could be confused with Eleonora's though considerably smaller, with faster wing-beats and shorter tail. Contrast between sooty-grey wing-coverts and paler silvery-grey flight-feathers above not seen in Eleonora's, which has plain dark upperparts (Fig. 65); underwing lacks pale line, pale patch on primaries and darker hind-wing of Eleonora's (Fig. 64), and normally has very characteristic silvery-grey primaries. Plumage differences of adult female (Figs. 67C, 68B), first-summer male (Figs. 67D, 68C) and juvenile (Figs. 67E, 68D) from Hobby already given; at distance where details become obscure, juvenile and adult females paler wing-coverts above, contrasting with darker flight-feathers are good distinguishing guide from dark uniform Hobby; also adult shows conspicuous pale upperhead. Juvenile further distinguished by yellower body and underwing-coverts and especially by distinct broad blackish-brown band along trailing-edge of whiter-barred flight-feathers; at closer range juvenile Red-footed also shows pale sides of neck connected with whitish half-collar on hind-neck, which is not a feature of juvenile Hobby. First-year male at certain stages of moult (see captions to Figs. 67 and 68) may superficially resemble male Merlin but differs in flight-silhouette by longer wings and tail. Red-footed habit of hovering also helps identification.

Fig. 67. Red-footed Falcon *Falso vespertinus*. Adult male (*above*), and (*opposite*) second-autumn male, first-summer male, juvenile and adult female from below.

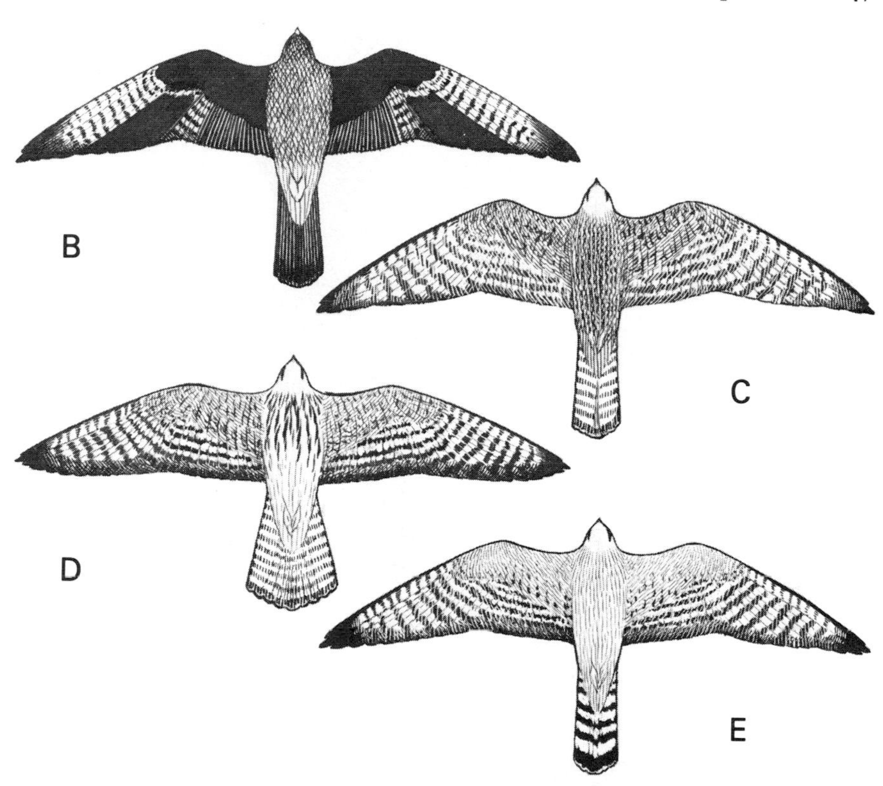

Fig. 67. **Red-footed Falcon** *Falco vespertinus*. Adult male, second-autumn male, first-summer male, juvenile and adult female from below. Adult male (A) is sooty or slate-grey on underparts, while lower belly, 'trousers', vent and undertail-coverts are deep dull-red, visible at close range, and undertail is blackish; underwing-coverts are slate-blackish, more or less contrasting with darkish grey flight-feathers and body below. Sub-adult male (B second-autumn, C 15 months old) in moult shows striking pattern of adult and juvenile flight-feathers, adult male-like underwing-coverts but paler ash-grey underparts, remains of moustachial streak and male-like, deep dull-red on belly-undertail-coverts; tail-feathers are slightly paler than adult's with blackish sub-terminal band (lacking in adult male). First-summer males show gradual change towards sub-adult plumage over a 3–5 month period: at first underparts are rusty- or rufous-brown with scattered thin blackish streaks, or more or less pale ash-grey underparts as in (C), throat is pale with short, diffuse, blackish moustachial streaks; flight-feathers show typical barring of juvenile, but dark trailing edge often inconspicuous (through bleaching); underwing-coverts also resemble those of juvenile i.e. densely dark-spotted. Tail-feathers normally largely juvenile but some birds moult early to show tail like (B). As moult advances underwing-coverts become dark slate as in adult male and birds resemble (A) but with barred flight-feathers and tail. Juvenile (D) has yellowish-white to pale buff underparts with streaking on breast becoming thinner on belly and short, dark, moustachial streaks on pale throat; rest of underparts unmarked. Strongly white and dark barred flight-feathers give effect of relatively pale ground colour which contrasts with broad blackish band on trailing edge of wing and blackish wing-tips. Underwing-coverts whitish to creamy-yellow with dense dark brown spots and bars, coverts looking darker than flight-feathers, but at distance whole underwing looks pale with dark trailing edge. Undertail densely barred creamy and blackish-brown (about nine narrow dark bars and broader sub-terminal one).

Adult female (E) is unstreaked pale rusty-beige on underparts and most underwing-coverts, paler, even creamy-white, towards belly; flight-feathers and primary coverts barred; remaining coverts sometimes lightly spotted. Flight-feathers like juvenile's but dark trailing edge less conspicuous or almost lacking; at distance wing-coverts look slightly paler than flight-feathers (reverse in juvenile). Undertail barred like juvenile's but smaller number of dark bars (7–8) are more widely spaced. First-summer female (not illustrated) very similar to adult. Legs are orange-red in males, orange in females and orange-yellow in juveniles.

Fig. 68. **Red-footed Falcon** *Falco vespertinus*. Adult male, adult female, first-summer male and juvenile (first-autumn) from above. Adult male (A) is dark slate-grey on head, upperparts and wing-coverts with contrasting silvery-grey flight-feathers and uniform blackish tail. Cere is red, bill dark with orange base. Adult female (B) is unstreaked rusty-brown to whitish-yellow (with bleaching) on top of head, paler on forehead and often whitish on hind-neck at close range; ear-coverts and sides of neck brownish-yellow or whitish and together with pale hind-neck often forming a half-collar. Eye-region, short, narrow eye-streak and moustachial streak blackish-brown. Upperparts and wing-coverts barred slate-grey and blackish, with paler slate-grey barring on lower back, rump and tail and rufous-brown wash on foremantle; flight-feathers only slightly darker brownish-slate and not contrasting very much with coverts, but wing-tips darker. At distance female looks fairly uniform slatish-grey above with distinctly paler head; cere is orange, bill dark with yellowish base. First-summer males vary with stage of moult and (C) (about one year old) shows bird which has moulted all juvenile wing-coverts and upperparts to dark slate-grey feathers, which show little contrast to dark brown, still-juvenile flight-feathers; central tail-feathers are new, being slate-grey, while remainder are juvenile (though a few birds

have moulted all juvenile feathers at this age). Hind-neck shows rusty-brown, forehead usually paler and thin, short, dark moustachial streaks contrast with whitish or pale golden brown cheeks. First-summer female (not illustrated) are very difficult to separate from adult females. Juveniles (D) resemble adult female but streaked 'cap' on head (darkest at rear) separated from rest of upperparts by whitish half-collar on hind-neck; dark on ear-coverts more pronounced than adult female. Upperparts and wing-coverts dark grey-brown, distinctly barred or edged paler brown on back and brownish-yellow on rump, tail-coverts and most wing-coverts which show pale line along tips of greater coverts; wing-coverts contrast with dark brownish flight-feathers more than in adult females and, at distance, approach that seen in female Kestrel. Uppertail is rusty or yellowish-brown with about ten black bars, wider sub-terminal bar and pale tip; central pair can be tinged slate-greyish.

Merlin
Falco columbarius
(plates 80, 81)

Merlins harassing Kestrel (below) and finches

SILHOUETTE Wing-span 52–69 cm. Male is smallest European raptor, but sexual dimorphism is considerable and female may overlap in length with other small falcons. Compact, with relatively long, square-cut tail and rather short, broad-based, pointed wings. Clearly shorter-winged than other falcons, with different wing-to-tail ratio (proportionately longer-tailed).

FLIGHT A bold and dashing little falcon with fast wing-beats interrupted by brief glides. More active, aggressive and determined in flight than larger species, soaring less and often flying low over the ground like a Sparrowhawk *Accipiter nisus*. When hunting, flight characteristically straight and purposeful on very rapid, short wing-beats with brief glides on closed wings, rather resembling the undulating flight of a large thrush (this applies especially to the small male). Capable of erratic twists and turns when pursuing prey.

IDENTIFICATION Male (Figs. 69A, 69B) easily distinguished from any other European falcon, being hardly bigger than Mistle Thrush *Turdus viscivorus*. Larger female (Figs. 69C, 69D) separated from Hobby and Red-footed Falcon and the two kestrels by more compact build, faster, more regular and determined wing-beats, and more direct and purposeful flight. Relatively shorter- and broader-winged than Hobby and Red-footed, without contrasting head markings. Shorter-tailed and with darker underwing than the two kestrels (and wings more pointed than in Kestrel); female's nearly uniform dark brown upperparts lack kestrels' contrast of red-brown and blackish. On the whole, female may often be more easily confused with Sparrowhawk than with another falcon, but Sparrowhawk has broader, more rounded wings which are also thicker with shorter hand when seen head-on.

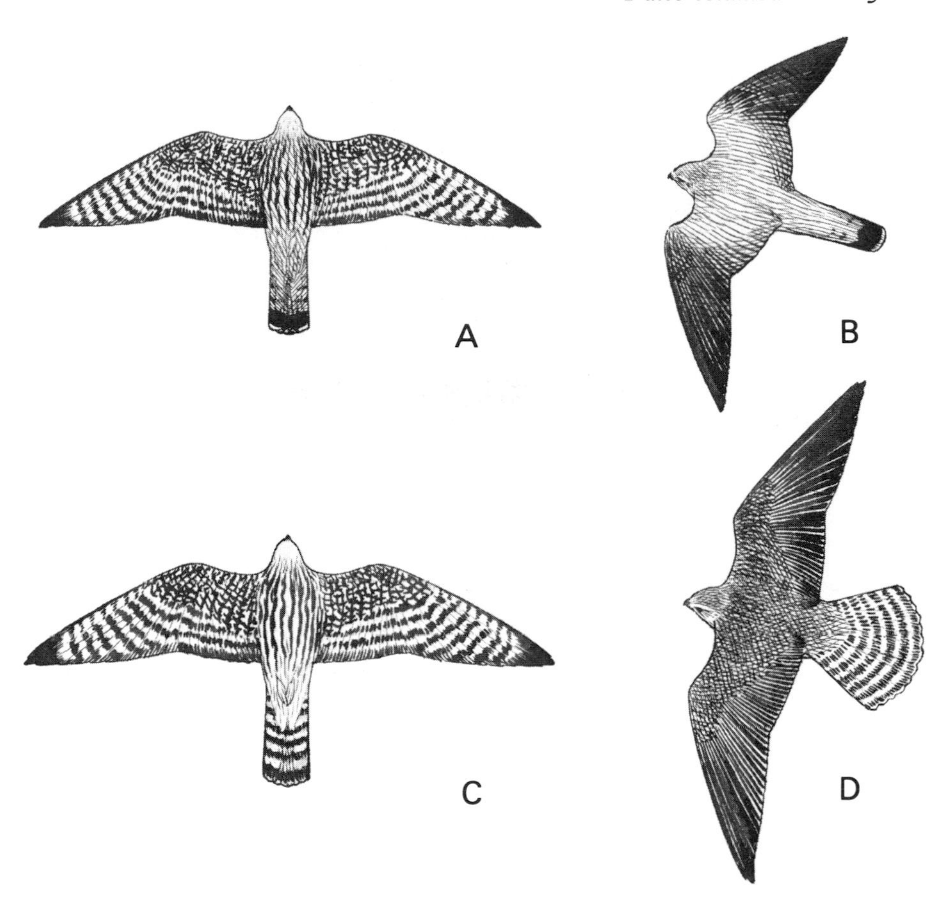

Fig. 69. **Merlin** *Falco columbarius*. Adult male and female from below and above. The sub-species *aesalon* is illustrated and described here; *subaesalon* of Iceland averages darker and larger. The underparts of adult males (69A, 69B) are beige to pale rusty, bleached paler in old plumage, with blackish-brown streaks which sometimes form drop-like spots on the flanks. The undertail is whitish with 3–4 narrow dark bars and a broad sub-terminal band with greyish-white tip, the underwing-coverts are whitish to buff, streaked and spotted reddish-brown, and the flight-feathers are whitish with dark brown barring. The upperhead is dark slate-grey, the nape rusty-brown, the back and wing-coverts slate-grey, and the uppertail pale slate, with a broad dark sub-terminal band and a narrow white terminal line (bars on the inner feathers are distinct only when the tail is fully spread). The greyish-black flight-feathers contrast with the paler wing-coverts. The head markings (pale forehead and supercilium, greyish or rufous sides and faint moustachial stripe) are visible only at close range. The underparts of adult females (69C, 69D) are whitish to creamy-white, heavily streaked dark reddish-brown and more spotted on the flanks. The underwing resembles the male's with coverts more reddish-brown, and the undertail is barred dark brown and whitish, with a wider dark sub-terminal band and a pale terminal line. The upperparts are largely brown, sometimes with a little white on the nape and a greyish wash on the back and rump, and rusty edges to the mantle and shoulders. The uppertail is dark brown with narrow whitish, but conspicuous, bars. The flight-feathers are dark brown above, with buff spots visible at close quarters. Inconspicuous head markings like the male's, but with the sides whitish streaked brown, are visible at close range. The legs and cere of adults are yellow, the bill bluish tipped darker. Juveniles (not shown) are hardly separable from adult females in the field, except for darker brown upperparts without a greyish wash (though this is seen in some males), and more brownish-white underparts, streaked darker brown. Young females often show darker cheeks and broader streaking below than young males. The cere is bluish, the bill and legs as in adults.

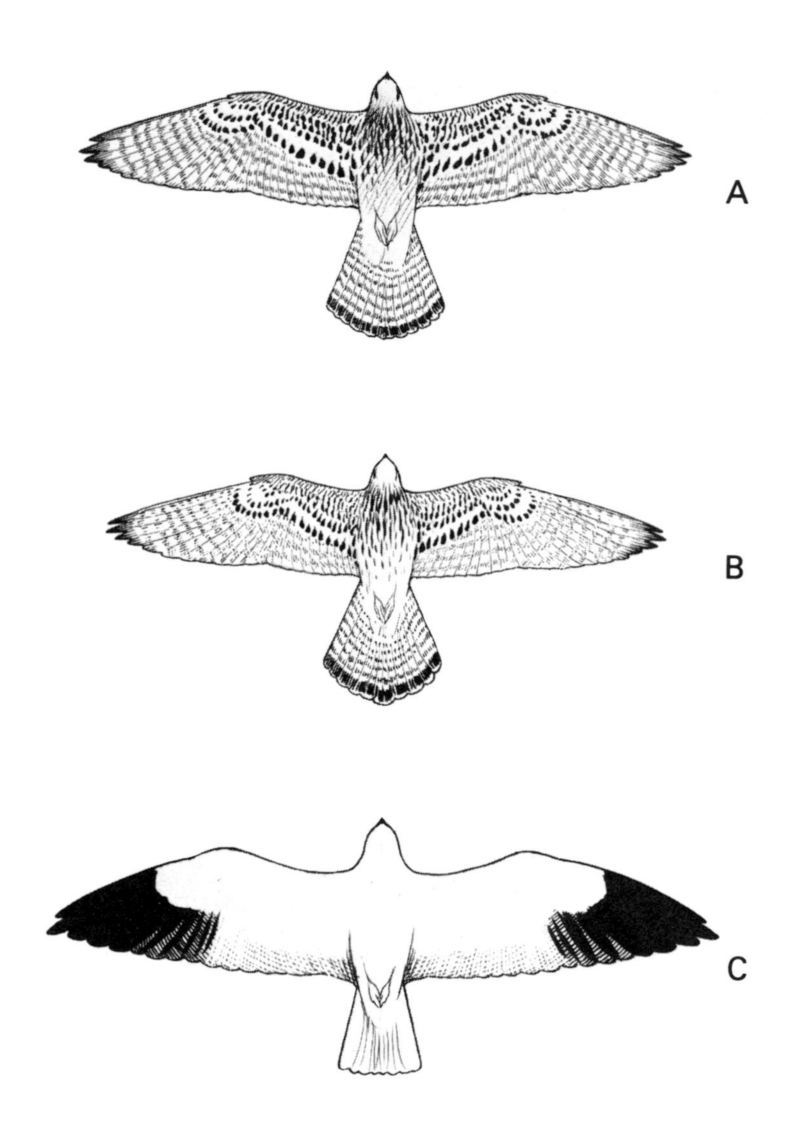

Fig. 70. Typical undersides of female Kestrel, female Lesser Kestrel and adult Black-winged Kite.

Fig. 70A. **Kestrel** *Falco tinnunculus*. Adult female shown. Small to medium-sized falcon with rather long pointed wings and long tail. Habitual hovering is a ready aid to identification. The male is easily identified by black spotted chestnut upperparts coupled with blue-grey head and tail; the latter with black subterminal band. The buff underparts are spotted black and there is an ill-defined black moustachial stripe. The female is dull chestnut-brown barred with black above, and buff below streaked black. It is very similar to female Lesser Kestrel (which see) and great care should be taken in identifying either. Fairly common in wide range of habitats throughout Europe (except Arctic Russia), north Africa, Turkey and Middle East, but some southwards migration in winter; summer visitor only in most of Fenno-Scandia and Russia.

Fig. 70B. **Lesser Kestrel** *Falco naumanni*. Adult female shown. Similar in structure to Kestrel though of slighter, more dainty build and tail more wedge-shaped. A colonial nesting raptor and on migration, too, can often be seen in small parties (but Kestels *F. tinnunculus*, particularly in southern Europe, sometimes nest in small colonies). The male is easily identified by its unspotted chestnut upperparts, with blue-grey greater coverts, pale blue-grey head (lacking moustachial stripe) and tail, the latter with a black sub-terminal band. The underparts are white, tinged rufous on the body with sparse darker spotting on the coverts, and dark wing tips. The female is very similar to female Kestrel (for details see text).

Breeds colonially in S. Portugal, most of Spain (except the extreme north), Mediterranean France, Sardinia, Sicily, most of Italy, S. Austria, the Balkans, Romania and S. Russia, also Cyprus, Turkey, Middle East and extreme N.W. Africa. Migratory, arriving mainly in April and departing in August–September. Winters in Africa south of the Sahara. In summer occasionally wanders north of breeding range in Europe.

Fig. 70C. **Black-winged Kite** *Elanus caeruleus*. Adult shown. Very slightly larger than Kestrel, to which it is somewhat similar in shape except for longer arm, more protruding owl-like head and shorter very faintly forked tail which is square cut when spread. Frequently hovers; glides with wings raised in an almost harrier-like manner. Plumage is unmistakable, being basically ash-grey and white with black primaries below and black shoulder patches on the upperwing.

In Europe occurs as a breeding bird only in Portugal and S. Spain (where rare) in open country with scattered trees or woodland. Stragglers are recorded regularly elsewhere in S. Europe, north to France and Germany. Nests also in N.W. Africa and Egypt, and widespread in tropical Africa, S. Asia and East Indies.

Kestrel
Falco tinnunculus
(plates 82, 83)

SILHOUETTE Wing-span 68–82 cm. A small to medium-sized falcon with relatively long pointed wings and rather long tail. It is heavier in build than the closely related Lesser Kestrel with slightly broader wings and less slender tail, rounded at the tip rather than wedge shaped as in many Lessers. The female is slightly larger than the male but size is usually only obvious when the two are seen together.

FLIGHT Open flight is a series of fast, shallow loose wing beats irregularly interspersed with glides and occasional twists and tumbles. Most frequently seen hovering over open country in search of prey and on such occasions wings are outspread and beat rapidly but shallowly and the tail fanned. Depending on weather conditions the wings may be held motionless as the bird floats in an upcurrent. The body angle of the hovering Kestrel varies between almost flat or at an angle of 45°. Sometimes soars with tips of wings slightly spread and tail fully fanned and on such occasions may resemble Sparrowhawk.

IDENTIFICATION The most similar species to the Kestrel, both in shape and plumage, is the Lesser Kestrel. Male Kestrel is told from male Lesser by black spots on chestnut upperparts and wing-coverts (unspotted and more extensive red in male Lesser, but beware, old male Kestrels on rare occasions have black spots much reduced or almost lacking); by chestnut, not largely blue-grey, greater coverts, innermost secondaries and largest scapulars. From below Kestrel has less whitish underwings which have generally more dark-barred flight-feathers, less dark wing-tips and trailing edge of underwing and generally more spotted underwing-coverts which contrast less with more streaked underparts. Diffuse dark moustachial streak of male Kestrel almost or quite absent in male Lesser.

Females are extremely difficult to distinguish outside nesting area (where approachable), the most valid character being colour of claws—blackish in Kestrel, whitish in Lesser. Experienced eye may detect slight differences in build, jizz, shape of tail, wing-tip formulae, strength of moustachial streak, streaking on upperparts, spotting on underwing-coverts and barring on flight-feathers.

The Kestrel is usually seen singly or in pairs whereas Lesser Kestrel is generally a colonial species, though caution should be taken as Kestrels are prone to breed in loose colonies of 2 or 3 pairs (occasionally more) in Southern Europe.

The Kestrel and Black-winged Kite are the only habitual hoverers amongst the smaller birds of prey. In silhouette these two species are quite different—the Kestrel having a smaller less noticeable head (this protrudes and is owl-like in Black-winged Kite) and longer tail.

The Lesser Kestrel also hovers but much less frequently, as does the Red-footed Falcon but this latter species as well as being quite different in plumage has a slimmer build and shorter tail.

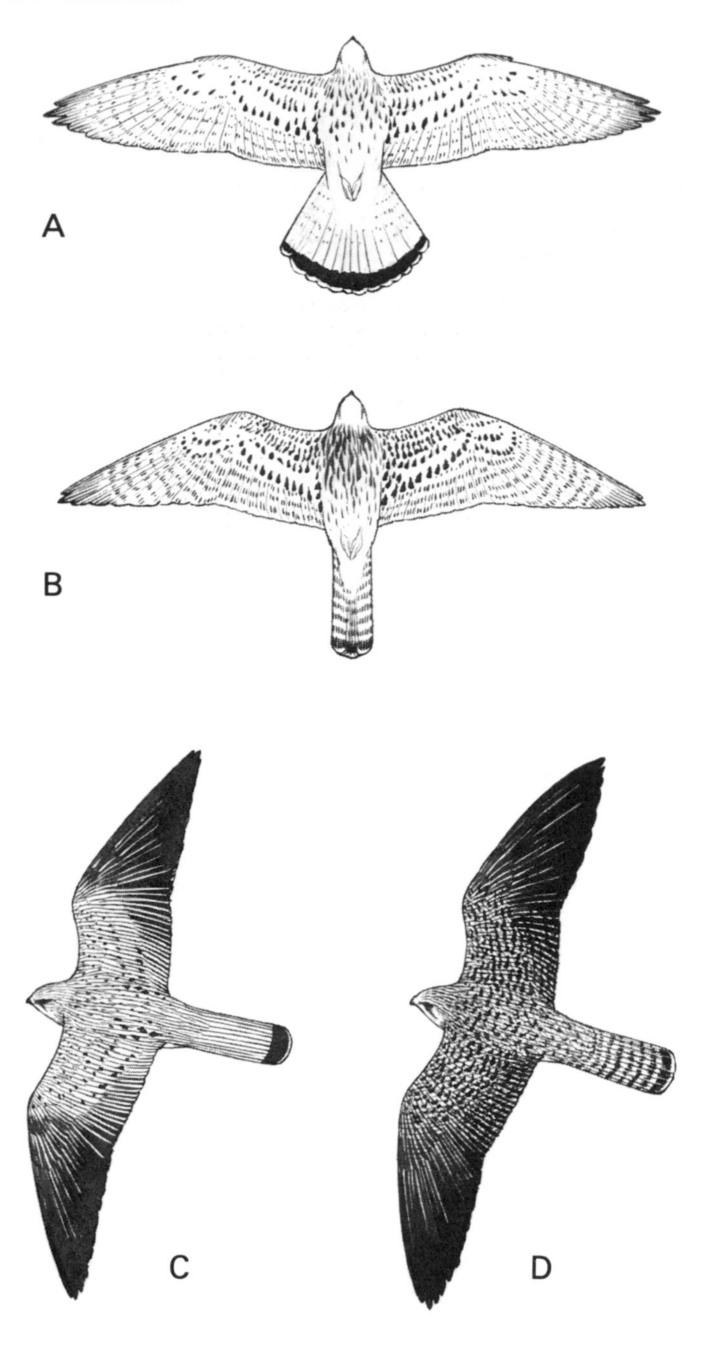

Fig. 71. Kestrels. Adult male (A) and adult female (B) from below; adult male (C) and adult female (D) from above.

Fig. 71. **Kestrel** *Falco tinnunculus*. Adult male and adult female from below, adult male (left) and adult female from above. The male from below is buff varying in shade and lightly spotted with black on the breast and flanks with a few spots on the belly. The underwing-coverts are buffish-white spotted with black, the largest spots being on the greater-coverts. The primaries and secondaries are whitish barred with grey and the wing-tips are darker. The tail is pale grey with a fairly broad black sub-terminal band, a narrow white terminal band and very indistinct barring on rest of tail which is rarely seen in flight. The female is buffish-cream below streaked blackish on the body—more heavily than male, and spotted on the underwing-coverts. The primaries and secondaries are clearly barred and the barred tail has pale tip and broad blackish-brown sub-terminal band. Juvenile (not shown) is hardly distinguishable from female in field.

The male from above has blue-grey crown, nape and hind neck with fine black streaks and ill-defined black moustachial streak. The back and upperwing-coverts are chestnut, spotted black and the primaries and primary coverts blackish-brown. Secondaries as primaries but progressively becoming chestnut with black flecking towards the body. Lower back, rump and tail blue-grey with a white tip to the broad black sub-terminal band on tail.

Adult female from above is dull chestnut-brown heavily barred blackish-brown and spotted or barred on back and rump. The upper head is chestnut-brown, streaked black and sometimes has bluish tinge. The flight-feathers are blackish, gradually becoming rufous, barred black on the inner secondaries. Uppertail-coverts dirty blue-grey with diffuse dark barring (seldom grey-brown). Tail, which appears slightly paler, is rufous-brown, sometimes washed greyish, narrowly barred blackish-brown and with wider sub-terminal band and whitish tip. Some females are tinged so blue-grey on head and uppertail that they can be easily taken for a male at distance, but dark barred tail, at least when spread, differs from males. Juvenile (not illustrated) is like adult female above but tail-coverts are brown, seldom tinged grey; uppertail is like female's generally (often greyer in young males) with dark barring more extensive across feathers, but of doubtful use in field. All birds have dark brown eyes and yellow cere, orbital ring and legs.

Lesser Kestrel
Falco naumanni
(plates 84, 85)

SILHOUETTE Wing-span 60–74 cm. Similar to Kestrel, but slimmer with slightly narrower wings and little shorter and more slender tail. The latter is frequently wedge-shaped caused by the central two tail feathers being more elongated, a feature that is usually more apparent in males than females, but Kestrel sometimes shows this, too.

FLIGHT Similar to Kestrel but tendency for wing-beats to appear lighter and shallower giving more frequently a 'winnowing' effect. Though it hovers regularly, it does so less persistently and less often than Kestrel, spending more time catching insects.

IDENTIFICATION The male Lesser Kestrel resembles the male Kestrel (which see) but should be readily identified by its cleaner whiter underparts, with white under wings contrasting with creamy-buff body, and black wing-tips; from above, lack of black spotting on chestnut back and coverts, and blue-grey greater coverts are diagnostic. The male Lesser Kestrel also lacks a moustachial stripe. Compared with female Kestrel, female Lesser has slightly whiter, less barred flight-feathers below, sometimes less densely and more finely-spotted underwing-coverts, slightly less developed dark moustachial streak, on average slightly thinner and neater streaking below and a little more grey on uppertail-coverts. Despite these differences, which are slight and inconsistent, identification of females is very difficult outside nesting area (where approachable and calling), if white claws and wing-tip formulae cannot be seen, the latter being visible on photographs, narrower wings and slightly shorter, narrower and frequently more pointed tip of tail are dubious characters. Although the Lesser Kestrel is a colonial species both in breeding and on migration, Kestrels, particularly in the southern part of their European range, occasionally nest in small colonies. So 'the rule', if it's on its own or in a pair it's a Kestrel, if it is in a colony it's a Lesser Kestrel, should be treated with extreme caution.

The male Levant Sparrowhawk may superficially resemble the male Lesser Kestrel at a distance, but whiter body does not contrast with underwings and tail is barred throughout, not pale grey, with only a distinct black sub-terminal band of male Lesser Kestrel.

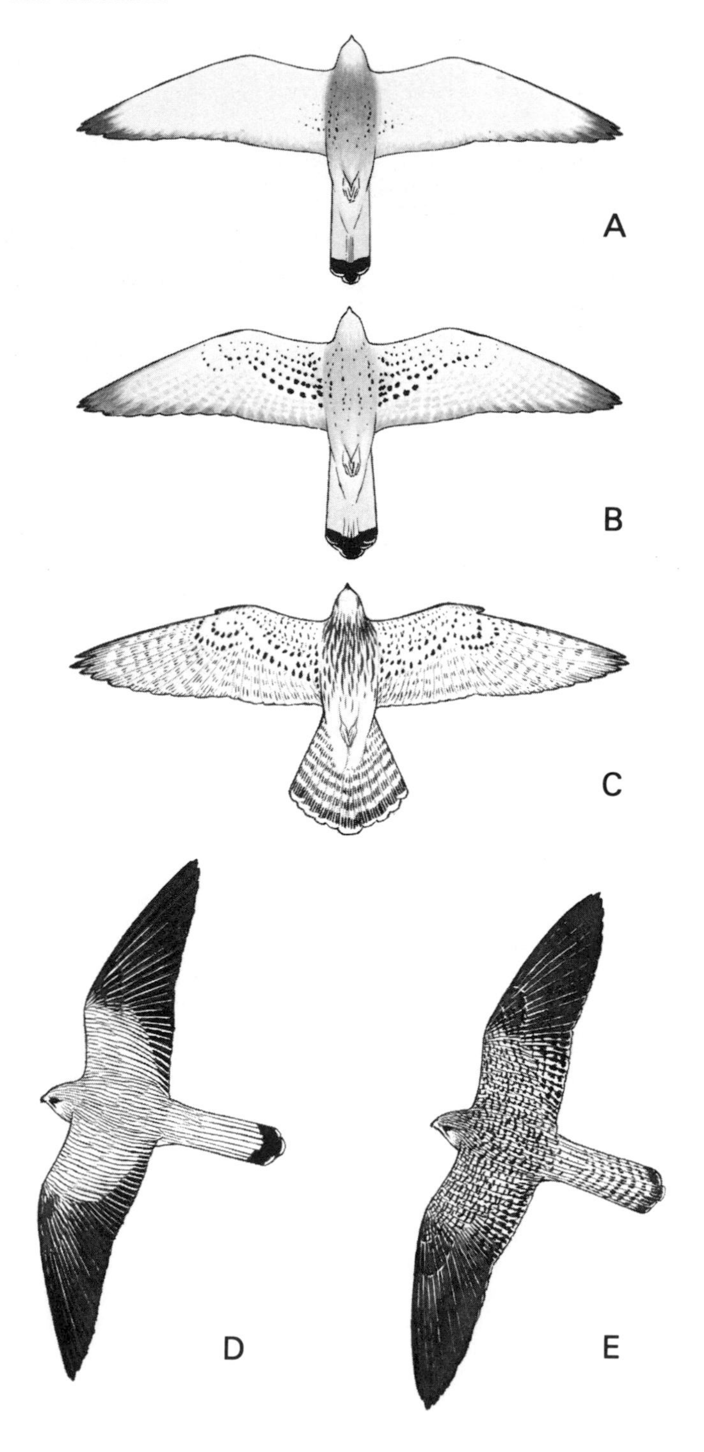

Fig. 72. Lesser Kestrel. Adult male (A and B) and adult female (C) from below; adult male (D) and adult female (E) from above.

Fig. 72. **Lesser Kestrel** *Falco naumanni*. Adult males and female from below and adult male and female from above. The underparts of male (A and B) are variably but rather intensely and uniformly coloured creamy-buff to deep reddish-cream with sparse, small black spots on breast and flanks, which can be reduced or almost absent; paler throat and lower belly to undertail-coverts. Underwing-coverts are creamy-white, either densely spotted (like male Kestrel) or almost completely unspotted. Flight-feathers are white, very faintly and thinly barred greyish, but with dark grey wing-tip and tips of most primaries, in some including whole of trailing edge of wing. Undertail is very pale bluish-grey or whitish-grey with broad black sub-terminal band, which appears triangular in shape when tail is closed (central feathers projecting a little behind other tail-feathers); tip of tail white. At distance it shows very white underwings with contrasting darker wing-tip and, in some, trailing edge of primaries, whole underwing contrasting with darker body (less apparent in some) and sparse spots on underparts 'disappear' at distance.

Female (C) is creamy-buff below, streaked blackish, generally fairly thinly though denser on upperbreast. Underwing-coverts are buffish-white, densely spotted blackish; in some, spots are confined largely to greater and median coverts. Flight-feathers are more clearly barred grey than in male (but generally less than in female Kestrel); wing-tip little darker. Undertail is buffish-white, narrowly barred darker and with broad, black, sub-terminal band and whitish tip. Juvenile (not shown) is very like adult female, though flight-feathers below are barred more strongly (like Kestrel). Bare parts are as in Kestrel (but claws whitish, not blackish).

From above, male (D) has unspotted, rich reddish-chestnut back and usually lesser and median wing-coverts, contrasting much with blackish hand (primaries, their coverts and outer secondaries). Most greater secondary-coverts, usually innermost secondaries and largest scapulars are blue-grey (chestnut in Kestrel); rarely all median and many of lesser coverts are also blue-grey. Head and neck blue-grey, at close range with darker shaded front of cheeks; rump, tail-coverts and unbarred uppertail pale bluish-grey, latter with broad black sub-terminal band and white tip.

Female Lesser (E) greatly resembles female Kestrel, being chestnut-brown above, narrowly barred blackish all over, with blackish-brown primaries and dark brown secondaries with pale barring. Brownish top of head is finely dark-streaked, forehead and cheeks paler, latter in front bordered by short, indistinct, dark grey-brown moustachial streak. Rump and uppertail-coverts often washed pale greyish; uppertail is chestnut-buff with narrow, blackish bars, broader sub-terminal band and white tip; tail-feathers regularly washed pale bluish-grey. Juvenile (not shown) closely resembles adult female and not safely distinguishable in field.

Black-winged Kite
Elanus caeruleus
(plates 86–88)

SILHOUETTE Wing-span 74 cm. A long-pointed-winged and, not least, long-armed bird of prey, a little larger than a Kestrel. Head protrudes noticeably and is broad and owl-like. The tail is fairly short and narrow with a hardly noticeable fork when closed but almost square cut when open. Tail length a little shorter than width of wings.

FLIGHT The flight is owl-like with soft fast wing beats. Hovers frequently and persistently, often in same spot for minutes, with the wings quite outstretched and the beats flat and fast. When gliding wings are raised like a harrier, but with hand more level, thus recalling miniature Rough-legged Buzzard. Wings are clearly angled when gliding, with carpal joint protruding as much as head, and hand angled backwards. Soars on wings held forward with lightly curved leading edges of wing, most noticeable on hand; trailing edge is straighter though not at right angles to body. Wings clearly raised, hand level.

IDENTIFICATION If a view is obtained in which the plumage details can be clearly observed there should be no problem in identifying this species. The falcon size, basically white and grey plumage with black wing patches and (below) black primaries are characteristic, though at a distance, if size cannot be judged, one must take care not to confuse with any of the pale grey and white male harriers, particularly Pallid Harrier, as the Black-winged Kite when gliding and soaring holds wings in a similar position to harrier.

If seen only in silhouette, the much protruding owl-like head, giving the impression that the wings have been set too far back on the body, should readily distinguish this species from any of the falcons particularly the Kestrel which also has a similar manner of hovering in search of prey.

Fig. 73. Typical head-on profile of soaring or gliding Black-winged Kite.

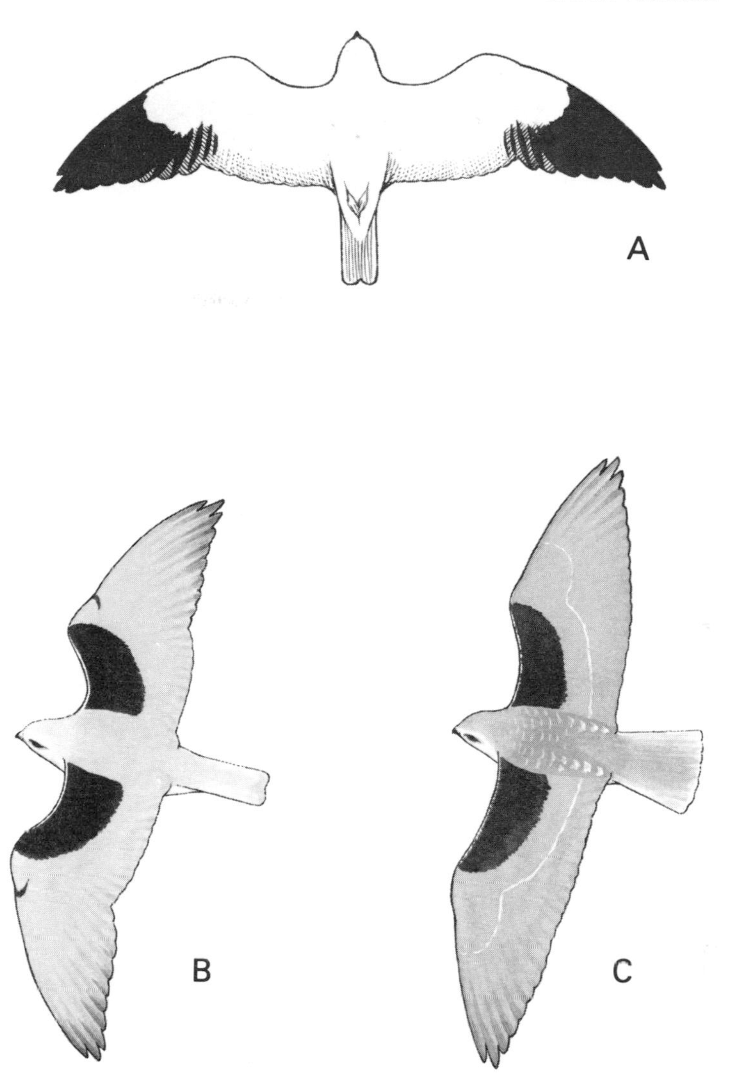

Fig. 74. **Black-winged Kite** *Elanus caeruleus*. Adult from below, adult and juvenile from above. Adult from below (A) is all white, except for blackish primaries and secondaries which are sometimes washed greyish, and very slight greyish wash on sides of breast. From above (B) has whitish head (white forehead and eyebrow) and pale ash-grey upperparts and upperwings grading into greyish-white uppertail, which looks white when spread; wing-tip darker grey, but most conspicuous mark is black median and lesser coverts; small black spot on greater primary coverts and white leading-edge of coverts. The bright red eye is almost entirely surrounded by dusky black, which gives head a fulmarine appearance.

Juvenile resembles adult below, though breast and flanks are washed yellowish-brown and usually have fine dark brown streaks and spots. From above (C) upperparts are brownish with narrow whitish edgings on back (soon lost through abrasion). Upperwing dark brownish-grey with white tips to greater coverts and, like adult, conspicuous black fore-coverts. Uppertail brownish-grey tipped whitish. Eye spot as in adult and whitish eyebrow. Eye is variable olive-grey, yellowish, red-brown or orange. This plumage lasts only about 3½ months when moult to adult plumage begins. First-year birds have fine whitish edges above and breast is washed greyish, but at end of moult are inseparable from adults.

7 The Accipiters

This can be quite a tricky group. Small male Goshawks are sometimes mistaken for large female Sparrowhawks and the latter can look similar to the immatures and females of the Levant Sparrowhawk.

All three are generally fast-moving species on the wing and prone to hunting in woodland rides, along hedgerows and amongst scattered trees and views are often brief. They are wood or forest nesters.

The Sparrowhawk and Goshawk are fairly widely distributed in Europe, though commoner in the northern parts; both partially migrate. The Levant Sparrowhawk, found only in S.E. Europe and Turkey east to Iran, is a true migrant; somewhat elusive in its breeding areas but common on migration at the Bosphorus in Turkey and various parts of the 'raptor route' down through the Middle East to Africa.

The Goshawk and Sparrowhawk are usually seen singly on migration, but the pale-plumaged Levant travels in flocks sometimes 500 or more strong.

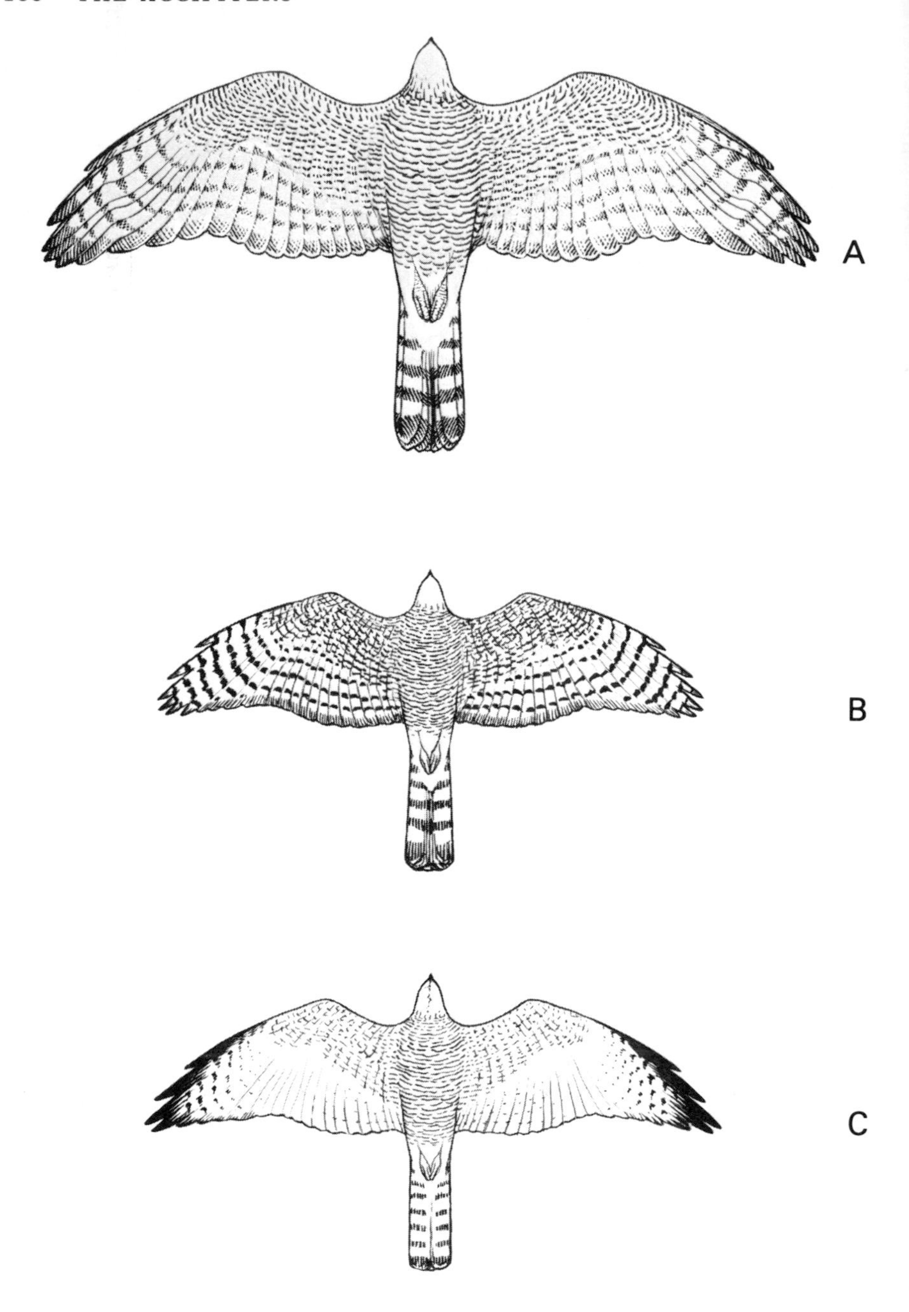

Fig. 75. Typical undersides of large female Goshawk, female Sparrowhawk and female Levant Sparrowhawk.

Fig. 75A. **Goshawk** *Accipiter gentilis*. Large adult female shown. The largest raptor in this group though there is much size difference between the smaller male and females and also variation between individuals of the same sex. Females are practically 'buzzard-sized' but with shorter wings, longer tails and more protruding head, giving at times an appearance not unlike a Honey Buzzard. Males are often little larger than large female Sparrowhawks. The upperparts of both are dark brown, tinged grey and the underparts and underwing-coverts whitish closely barred with brown. Adults show a conspicuous superciliary stripe.

In Europe three races occur—the nominate *A. g. gentilis* in Norway, Sweden, Finland, British Isles (very rare), south to the Mediterranean and across Europe to Turkey. In the extreme north of Sweden and Finland is the larger paler *A. g. buteoides* which winters further south in Fenno-Scandia and central Russia, occasionally coming south to Holland, Germany and Hungary. The race *A. g. arrigonii* occurring in Corsica and Sardinia is smaller and darker.

Fig. 75B. **Sparrowhawk** *Accipiter nisus*. Adult female shown. Female is only a little smaller than many male Goshawks and care is needed to distinguish between them. The Sparrowhawk however is proportionally shorter winged and has square-cut, not rounded tail of the Goshawk. Plumage of female very similar to Goshawk, though more finely barred. Male Sparrowhawk is a very small bird of prey with short wings, long tail and orange barred underparts. Flight is swift, being a series of rapid wing-beats interspersed with glides on flat wings. Nominate race breeds in Britain, Ireland, central Europe south from about 69°N, Sicily and northern Turkey; both sedentary and migratory, some birds moving south in winter, especially from northern Europe. A similar race (*punicus*) is resident in north-west Africa; and the smaller, darker and more densely barred *wolterstorffi* is resident in Corsica and Sardinia.

Fig. 75C. **Levant Sparrowhawk** *Accipiter brevipes*. Adult female shown. Similar in shape to Sparrowhawk but with slightly narrower, more pointed wings, giving the bird a falcon-like appearance; also shorter tail. Adult male is unmistakable with white underparts suffused on the body and underwing coverts with vinous buff, and black wing-tips. Upperparts are pale blue-grey giving the bird a 'dove-like' appearance. Female is barred below, but is paler than female Sparrowhawk; black wing tips and more dark bars on tail than in Sparrowhawk should aid identification. Immature, which also has dark wing-tips, is brown above with heavily checkered underparts formed by brown drop-like spots on a whitish body.

A migrant reaching S.E. Europe in April and May and departing in September. Breeds in S.E. Europe in the Balkans (except N. Yugoslavia), eastern Romania, the Ukraine and W. Turkey.

Goshawk
Accipiter gentilis
(plates 89, 90)

SILHOUETTE Wing-span 96–127 cm. Male resembles large female Sparrow-hawk but has proportionally slightly longer, broader arm and correspondingly shorter, more pointed hand, generally giving more pronounced curved trailing-edge to wings; tail a little shorter proportionally than Sparrowhawk's with more rounded corners; comparatively more protruding head, thicker neck and heavier, deeper body. These differences are even more pronounced in considerably larger female.

FLIGHT Heavier and more powerful flight than Sparrowhawk with heavier, slower and flatter wing-beats, interspersed with short or long (particularly on migration) straight glides on flat, rather pointed wings which have pronounced S-curved trailing edge. Frequently soars, particularly over breeding grounds and there is tendency for tail to be more fully spread than Sparrowhawk's, but it is not infrequent for Goshawks to soar with closed tails. Although capable of great speed, even smaller males appear much slower and heavier in flight than Sparrowhawks, less abruptly twisting and turning after prey, but beware Sparrowhawks' 'slow-motion' spring display flights when they may appear much larger than they really are.

IDENTIFICATION Buzzard-size wing-span of female rules out confusion with female Sparrowhawk (see also SILHOUETTE). Its longer tail and more pointed wings and typical *Accipiter* flight (see FLIGHT) rule out confusion with Buzzard. Harriers are much more slender, have longer wings which are raised in shallow V when soaring and gliding. Honey Buzzard is quite different with proportionally longer wings, narrower neck and less rounded wing-tip; plumage also differs considerably. Female Goshawk not infrequently confused with Gyr Falcon (even by experienced observers) which it approaches in wing-span; but Gyr has longer and much more pointed hand, and trailing-edge of wings is not S-curved; moreover tail markings of Goshawk with about four conspicuous broad bands is good guide (Gyr has many narrower bars).

Real danger of confusion with other species arises between male Goshawk and female Sparrowhawk where size difference not always apparent, though male Goshawk usually is clearly larger than a Crow *Corvus corone,* which female Sparrowhawk is not (wing-span of female Sparrowhawk up to 80 cm, male Goshawk down to 96 cm). Goshawk's longer arm and comparatively shorter, more pointed hand, more curved trailing edge of wing, shorter, more rounded tail (more square-cut or 'notched' in Sparrowhawk), heavier, slower wing beats and heavier, deeper body, distinguish it from female Sparrow-hawk, which sometimes looks conspicuously long-tailed. Plumages are rather similar, though barring of flight-feathers on underwing of Goshawk is less crisp and often looks 'washed-out', particularly in older birds; white undertail-coverts are usually more conspicuous and thicker-, more bulky-looking in Goshawk, whose dark brown eye-streak and upper ear-coverts together with darker top of head give it a more hooded appearance.

Juveniles and immatures are heavily chequered below (rather like large juvenile Levant Sparrowhawk) and lack white undertail-coverts and dark ear-coverts of adults. Apart from these features, identification points mentioned above apply, except for underwing barring on flight-feathers which is dark and neat in young birds.

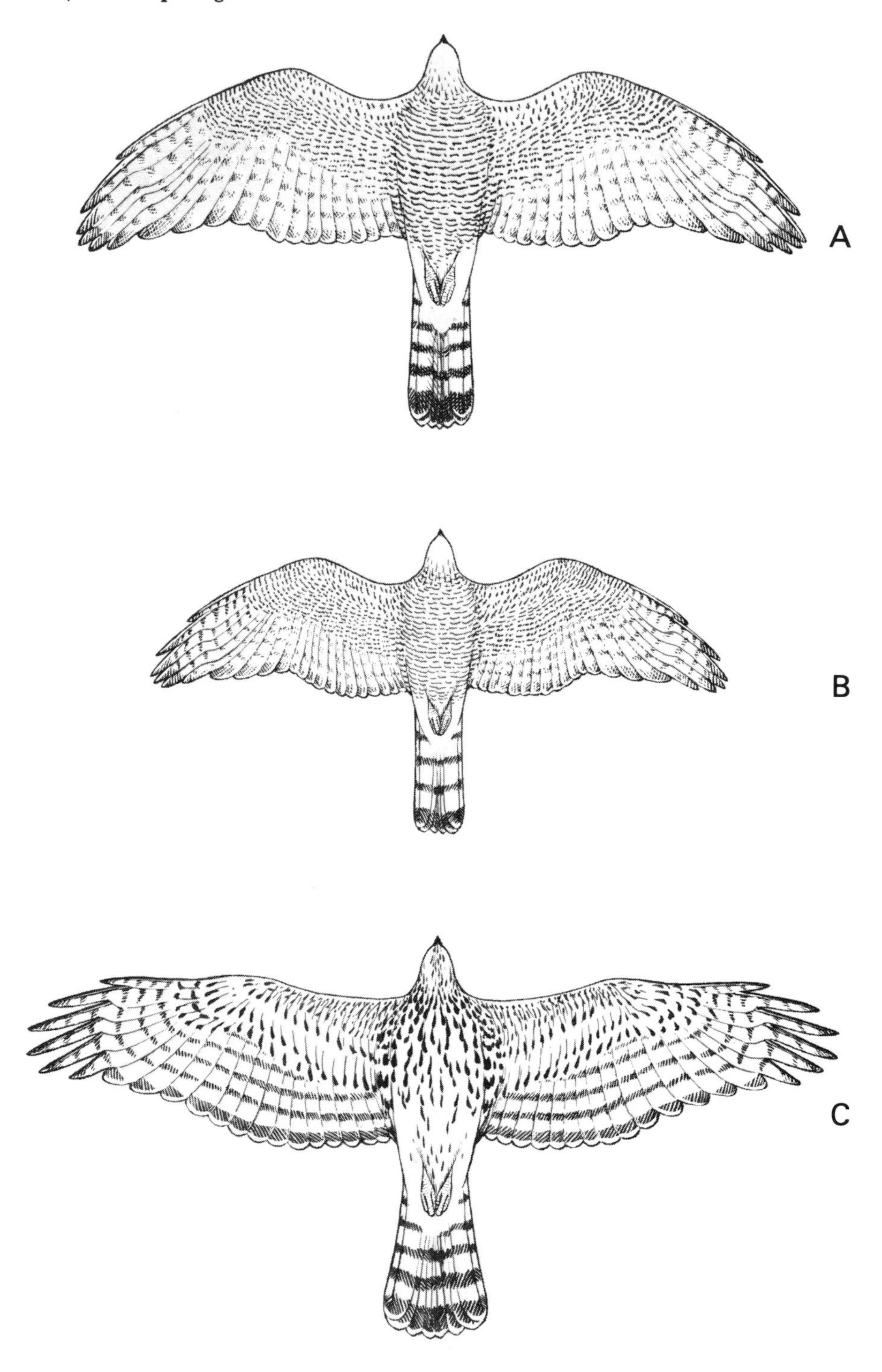

Fig. 76. Goshawk *Accipiter gentilis*. Adult female, adult male and juvenile female from below; adult female and juvenile male from above.

Fig. 76. **Goshawk** *Accipiter gentilis*. Adult female, adult male and juvenile female from below; adult female and juvenile male from above. Apart from difference in size, female (A) and male (B) are similar from below. Underparts are whitish, closely barred blackish-brown on body and underwing-coverts; flight-feathers are dark barred, less conspicuously than in juvenile and becoming progressively more 'washed-out'-looking as bird increases in age, and undertail-coverts can be, but are not always, conspicuously white. Undertail is greyish-white with four to five dark bars of which sub-terminal is broadest; tip whitish. Juvenile (C) has whole underparts buff or even rufous-buff, bleaching to paler yellowish-white and variably streaked with dark brown drop-like markings, but 'trousers', vent and undertail-coverts less streaked or unmarked. From above adult male (not shown) has top of head and eye-stripe blackish-brown giving hooded appearance and there is broad whitish supercilium and some white on hind-neck. Upperparts and wing-coverts are slate-grey or grey-brown, faintly tinged bluish, flight-feathers brownish, only slightly darker than coverts, and there is diffuse pale patch on primaries. Uppertail is grey-brown with about four blackish-brown bands and white tip. Large female (D) is similar above, though upperparts not so bluish. Juvenile (E) has dark brown upperparts and wing-coverts, most feathers broadly edged buff or rusty-buff and with pale line on tip of greater coverts and trailing edge of wing; at distance has buffish-brown appearance, particularly on wing-coverts. Top of head not so dark as adults, supercilium is diffuse or lacking as is dark eye-streak; sides of head buffish-brown streaked darker and there is pale patch on hind-neck. Uppertail as adult but dark bars more obvious and tip buffish and broader.

Birds from extreme N. Fenno-Scandia *A. g. buteoides* are pale and more blue-grey above (adults) and whiter below with finer or reduced barring. Juveniles of this race are paler above with some buffish feather edgings and often with visible whitish bases to feathers of back and wing-coverts giving mottled appearance; at distance these markings sometimes form palish patch on whole centre of forewing. Birds in southern part of their range in Europe are generally darker than typical examples of nominate race described above.

Sparrowhawk

Accipiter nisus

(plates 91–94)

SILHOUETTE Wing-span 60–80 cm. A small hawk with well-proportioned body; rather short, relatively broad and rounded wings (five free 'fingers') of which 6–7th primary are longest (useful in distinguishing Sparrowhawk from Levant Sparrowhawk in photographs) and rather long tail; latter is square-cut or slightly notched (not so rounded as in Goshawk). Unlike falcons, wings in open flight and in soaring are blunt-ended. Female slightly but usually noticeably larger than male which can often resemble a long-tailed Merlin.

FLIGHT Active flight is a series of rapid wing beats interspersed with short glides on flat wings. When hunting speed accelerates, wing beats becoming faster; flight is often low over ground, hiding behind cover before sudden attack, but may frequently be seen soaring high above ground before hunting starts; when soaring wings are held flat and slightly forward and tail usually fully closed, though it is often half-spread at onset of a period of soaring.

IDENTIFICATION Confusion can arise with Goshawk, Levant Sparrow-hawk and, less frequently, Kestrel and Merlin.

Large female Sparrowhawk is only a little smaller than male Goshawk and points of distinction are Sparrowhawk's proportionally shorter, less pointed and less curved wings which have relatively shorter arm and longer hand than Goshawk, its longer tail which is square-cut or slightly notched, its relatively smaller head, more slender neck and body which is not so bulky as in Goshawk; wing beats are also faster and lighter and glides usually shorter—a point that is particularly noticeable in the two species on migration.

Sparrowhawk can be distinguished from Levant at all ages by absence of dark wing tips (though occasionally juvenile Levant lacks this) and by less pointed, slightly broader wings (Levant more delicate and falcon-like; for wing-formula see Levant). Tail-barring also differs: Sparrowhawk has four to five bars visible in the field, compared to five to seven narrower bars in Levant (see plate 95). Differentiation between males is easy—the Levant being almost white on underwings with black tips.

Kestrels and Merlins have longer, much more pointed hand and different flight.

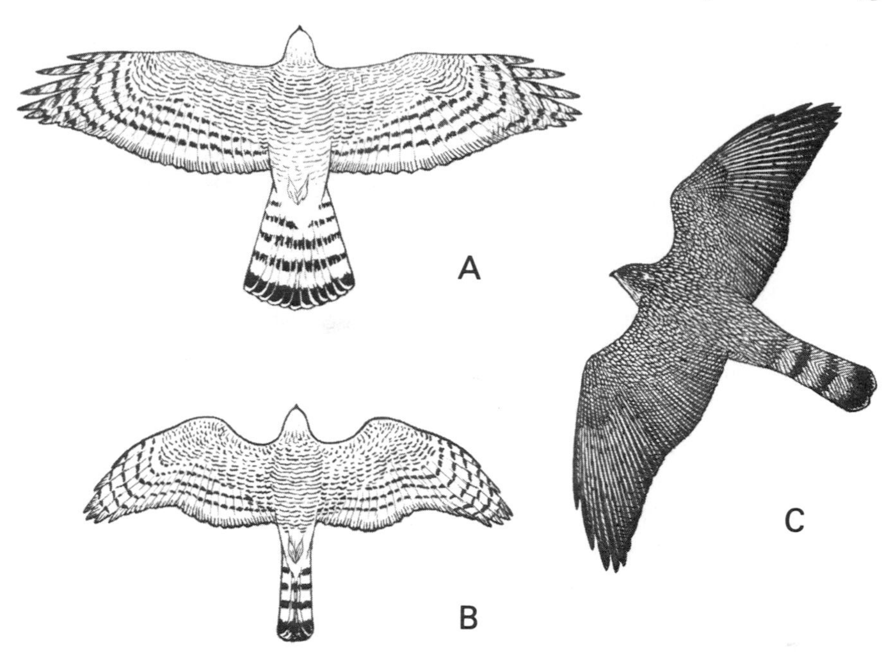

Fig. 77. **Sparrowhawk** *Accipiter nisus*. Adult female and male from below and adult from above. Underparts of adult male (B) are white, very closely barred with reddish-brown or orange; some have grey-brown barring but bars on sides of body are always reddish; some appear uniform deep orange at a distance; conspicuous white undertail-coverts usually unbarred; throat whitish and cheeks rusty-brown. Underwing-coverts creamy-white, closely barred dark brown, flight-feathers greyish-white, barred blackish brown, generally more conspicuously than in adult Goshawks. Greyish-white undertail has four to five dark brown bars of which sub-terminal is broadest; tip of tail white; Larger female (A) has stronger grey-brown or blackish-brown barring below and sometimes warm orangey barring on sides of body and only occasionally with largely reddish barring; cheeks and sides of neck often washed reddish; otherwise much like male. Juvenile from below (not shown) has, except for dark brown streaked sides of head, creamy-white underparts with more irregular, ragged and often broader rusty to dark brown barring, but chin and throat are streaky and in young females lower neck and upperbreast tend to be more streaky or spotted instead of barred.

Upperparts of adult male (C) usually dark slate or bluish-grey with head and mantle slightly darker (some males appear quite pale blue-grey above). Forehead is tinged rufous and there is often small whitish spot on the nape. Flight-feathers are dark grey-brown with slightly paler dark-barred diffuse patch on primaries. Uppertail is brownish with four to five broad dark bars and tip white. Female above is dark grey-brown, often with slate-grey cast, but not so bluish as in males. There is narrow white superciliary stripe (usually lacking entirely in male) bordered below by dark eye-stripe. Otherwise female much like male. At distance upper-wings look rather uniform with just slightly darker flight-feathers. Juvenile from above (not shown) is warm dark brown on upperparts and wing-coverts with pale rusty-brown feather edgings and a whitish nape spot. Identifying juveniles requires close observation.

Levant Sparrowhawk
Accipiter brevipes
(plates 95, 96)

SILHOUETTE Wing-span 64–80 cm. A more 'falcon-like' Accipiter than Sparrowhawk, due to slightly more slender, longer and more pointed wings, which are also less curved on trailing-edge (four free 'fingers' of which 7–8th primaries are longest); tail also a little shorter than Sparrowhawk's and more often rounded or pointed (not notched like many Sparrowhawks).

FLIGHT Similar to Sparrowhawk but wing-beats slightly slower and with less dashing flight; tendency for longer periods of soaring and gliding; much more likely to be seen soaring in tight flocks (often large) on migration or gliding closely together between thermals.

IDENTIFICATION Confusion can arise with Sparrowhawk and even male Lesser Kestrel. Male Levant is unmistakable provided a good view is obtained; its white underparts suffused on body and coverts with pinkish-buff and black wing-tips provide for quick identification, though due to its more pointed-winged appearance compared to other Accipiters, care should be taken with male Lesser Kestrel which is also very pale bird below. This latter species, however, is always readily identified, even at distance, by black terminal band on tail.

Female Levant can be distinguished from Sparrowhawk by more slender and more pointed wings, shorter and more rounded tip to tail, usually paler appearance below with darker wing-tips (which Sparrowhawk lacks). Also Levant has five or six narrow dark bars on undertail (even up to seven on outermost feathers) compared with four to five broader bars in Sparrow-hawk, a feature that is valid for all ages. In its juvenile and immature form Levant also shows dark wing-tips but less obviously than in adults; it has chequered underparts, the drop-like spots on the whitish underbody being quite unlike ragged barring of the underparts of young Sparrowhawk.

(*opposite*)

Fig. 78. **Levant Sparrowhawk** *Accipiter brevipes*. Adult female (A), adult male (B) and juvenile (C) from below; adult male (D) and juvenile (E) from above. Adult male below has white underparts and wing-coverts densely and very narrowly barred pinkish- or cinnamon-red, but in some the barring is washed-out on breast to an almost uniform colour; white throat has faint dark central streak. Flight-feathers are white, very faintly and narrowly barred greyish on bases; narrow diffuse dark trailing edge to secondaries which becomes broader and blacker on primary tips, contrasting greatly with white underwings; at distance barring below 'disappears' and whole under surface looks strikingly white with blackish wing-tips. Undertail is whitish with usually four to six bars visible in field but sometimes more on outer feathers. Adult female below has whitish underparts and coverts with broader and more brownish bars, which are thinner on coverts. All flight-feathers are clearly barred dark grey and are more Sparrowhawk-like, though narrower generally, but tips of outer primaries are blackish; at distance female is not so whitish-looking as male but darker wing-tips are discernible in good light. Undertail bars slightly broader than male's. Juvenile below has creamy-white underparts heavily marked dark or rusty-brown with longitudinal or heart-shaped spots; wing-coverts barred. White throat has clear blackish central stripe. Flight-feathers below are whitish, distinctly barred dark brown, generally narrower than in Sparrowhawk and extreme tips of outermost primaries are dark grey; at distance looks longitudinally streaked below, distinctly barred on flight-feathers and close to

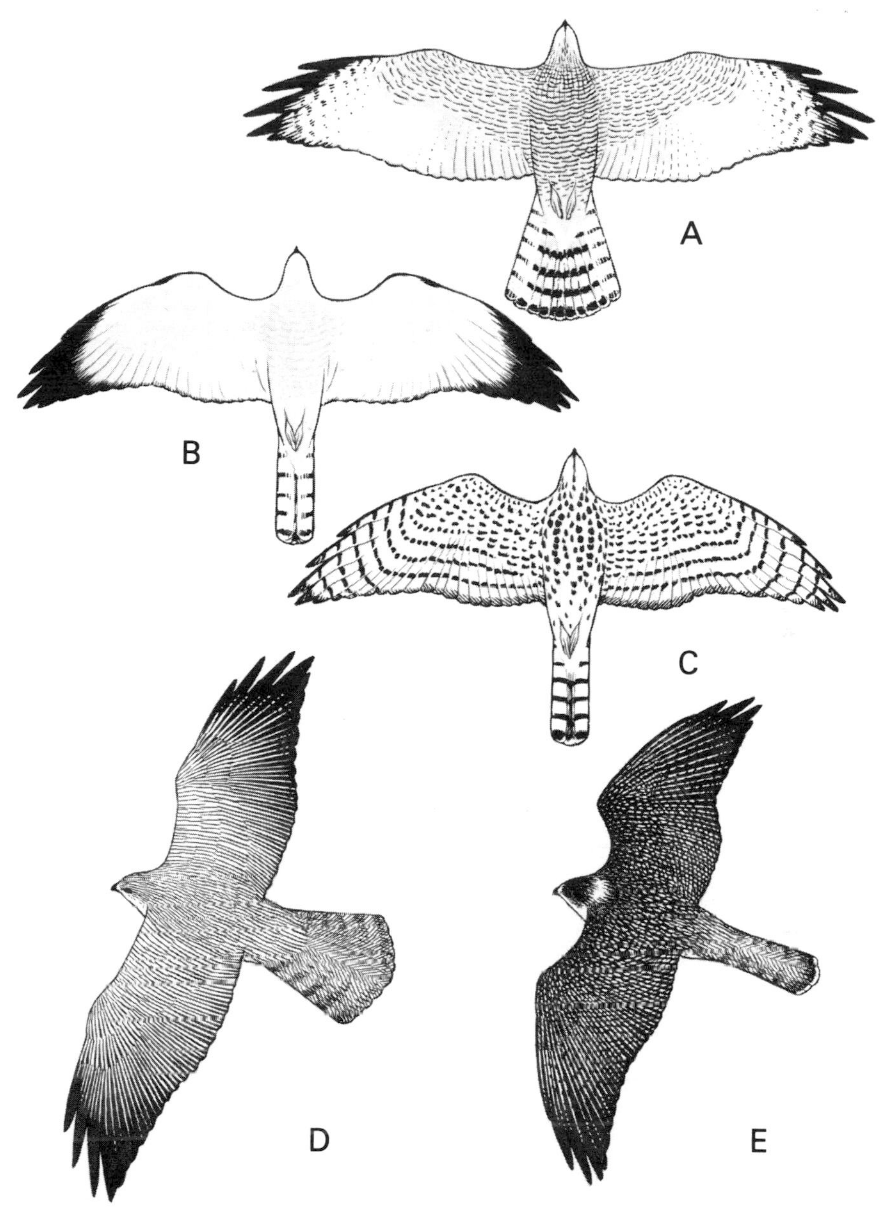

in good light usually shows darker tips to outer primaries. Undertail as adult but broader pale tip to tail.

Adult male from above is beautiful blue-grey with outer half of primaries blackish giving almost dove-like appearance. Ear-coverts grey with rusty-brown on sides of neck. Closed tail is blue-grey but outer feathers brownish-grey with four to seven dark bars, sub-terminal widest. Female is dark brownish-grey above with slightly darker wing-tips, but in some old birds almost as grey above as adult male. Closed tail dark brown, often with diffuse dark sub-terminal band, and even three to five faint bars at close range; outer feathers show strong barring. Cheeks, ear-coverts and sides of neck uniform grey-brown.

Juvenile dark brown above, palest on rump, darkest on mantle and almost blackish-brown on foremost coverts; hind-neck shows variable whitish spot. Uppertail resembles female's but barring more evident on central feathers.

8 Summary of the Legal Status of Birds of Prey in Europe

We are most grateful to the World Working Group on Birds of Prey for allowing us to summarise the findings of their survey which started in 1970 on the Legal Status of Birds of Prey in Europe. This we have since updated to take account of recent changes in legislation, the most notable of which has been the introduction of full protection for birds of prey in Malta in 1980.

As the 1970 survey also covered owls we have included them in this summary, though of course their identification is not treated in this book.

A *Countries affording partial protection to birds of prey:*

AUSTRIA

The situation is complicated by the varying degrees of protection afforded by the different states. All vultures, eagles, kites, falcons and owls are fully protected everywhere. *Accipiter gentilis, Buteo buteo, B. lagopus, Circus spp.* and *Accipiter nisus* are fully protected in some states but only partially in others. During the open season, which varies between states, a special permit is required for the hunting and keeping in captivity of birds of prey.

BULGARIA

All but four species of birds of prey and owls are protected by special law; the exceptions are *Buteo lagopus, Circus aeruginosus, Accipiter gentilis* and *A. nisus.*

CHANNEL ISLANDS

This small group of islands is not part of the United Kingdom and some islands have their own conservation laws. On the island of Alderney, all birds of prey are protected except *Accipiter nisus.* In Jersey and Guernsey all birds of prey are protected.

CZECHOSLOVAKIA

All birds of prey and owls are protected but *Accipiter gentilis, Buteo buteo* and *B. lagopus* may be killed in pheasant preserves.

DENMARK

All birds of prey and owls are protected, except for *Buteo buteo, Accipiter gentilis* and *A. nisus* which may be killed in certain circumstances (for example, where they are taking pheasants), if special dispensation has been given by the Ministry of Agriculture.

EAST GERMANY

The picture is complicated. Species that are known to be protected are the *Aquila* eagles, *Circaetus gallicus, Circus pygargus, C. cyaneus, Pandion haliaetus, Falco peregrinus,* Eagle Owl *Bubo bubo* and Pygmy Owl *Glaucidium passerinum.* The nests of *Haliaeetus albicilla* are given protection. *Accipiter gentilis, A. nisus, Buteo buteo* and *B. lagopus* are protected outside the hunting season.

FINLAND

All birds of prey and owls are totally protected except for *Buteo lagopus* in Lapland, *Accipiter gentilis* and *A. nisus* throughout the country, and Eagle Owl *Bubo bubo* outside the breeding season. These species are unprotected on the grounds of their alleged game damage which the national ornithological society is trying to refute.

GREECE

All birds of prey and owls are fully protected, except within breeding, hunting and wildlife reserves, where these birds are considered harmful. Ornithological societies are working to eliminate this provision so as to ensure total protection is afforded to all birds of prey.

HUNGARY

All birds of prey and owls are fully protected with the exception of *Buteo lagopus*. *Accipiter gentilis, A. nisus* and *Circus aeruginosus* are only protected during the breeding season, except that the *Accipiters* can be killed by game farms.

IRISH REPUBLIC

The Wild Birds Protection Act 1930 affords a degree of protection to birds of prey and owls. All species are given a close season between 1 March and 31 July, which can be adjusted by the Minister for Justice in any one county. It appears that the owner or occupier of land may kill some birds of prey, including *Buteo buteo, B. lagopus, Circus cyaneus, Accipiter nisus* and *Falco tinnunculus*. There are proposals for comprehensive wildlife legislation in the Republic of Ireland.

ITALY

There is no federal law, but provincial regulations related to protection of birds of prey and owls. The provinces which apply full protection are Val d'Aosta, Piemonte, Liguria, Lombardia, Veneto, Friuli-Venezia Guila, Emilia, Romagna, Toscana, Lazio, Marche, Molise, Trentino-Alto Adige and Sardegna. *Gypaetus barbatus* is the only species protected everywhere, though it is now possibly extinct in Italy. No species is protected in hunting preserves and restock and capture areas (which form less than 10% of the whole country, subject to the exceptions above); in these areas they may be shot at any time and by any means. However, with the exception of Eagle Owl *Bubo bubo*, Barn Owl *Tyto alba* and Little Owl *Athene noctua*, all owls are protected outside the hunting preserves; and, with the exception of *Milvus milvus* and *M. migrans*, the same applies to all diurnal birds of prey from 31 March to 30 August.

ROMANIA

All birds of prey and owls are protected except for *Circus aeruginosus, Accipiter gentilis* and *A. nisus*.

SWEDEN

All birds of prey and owls are fully protected throughout the whole country except for *Accipiter gentilis*. This species may be killed in areas where a game-rearing farm, poultry-farm or similar establishment is situated, in order to prevent damage to animals on the farm. No birds of prey or owls may be used as decoys.

SWITZERLAND

All birds of prey and owls are protected, but individuals of *Buteo buteo, Accipiter gentilis* and *A. nisus* causing serious damage may be killed under authorisation of the competent Cantonal authority.

WEST GERMANY

All birds of prey and owls are generally protected, but in some provinces *Buteo buteo* and *B. lagopus* may be hunted from 1 November to 28 February except in state-owned, restricted hunting areas. Pressure is being applied to have all birds of prey and owls protected throughout the Federal Republic.

YUGOSLAVIA

In Croatia all birds of prey are fully protected. However, *Accipiter gentilis, A. nisus* and *Circus aeruginosus* are not protected anywhere else, nor are *Aquila chrysaetos* (which may be shot from 1 September to 12 December), *Buteo buteo, B. lagopus* and *Falco subbuteo* in Slovenia.

B *Countries affording full protection to birds of prey*

The countries which afford protection to all birds of prey and owls are:

BELGIUM
CYPRUS
FRANCE
GIBRALTAR
LIECHTENSTEIN
LUXEMBOURG
MALTA
THE NETHERLANDS

NORWAY (although *Aquila chrysaetos, Haliaeetus albicilla* and *Accipiter gentilis* may be killed under special dispensation)

POLAND
PORTUGAL
SPAIN
TURKEY
UNITED KINGDOM

It should be noted that this summary is intended as no more than a general outline. For fuller details and verification it is essential to contact the appropriate ministry in the country concerned. However, as far as it has been possible to check, this outline is correct as at December 1980.

Index of Scientific Names

(Numerals in italics are plate numbers)

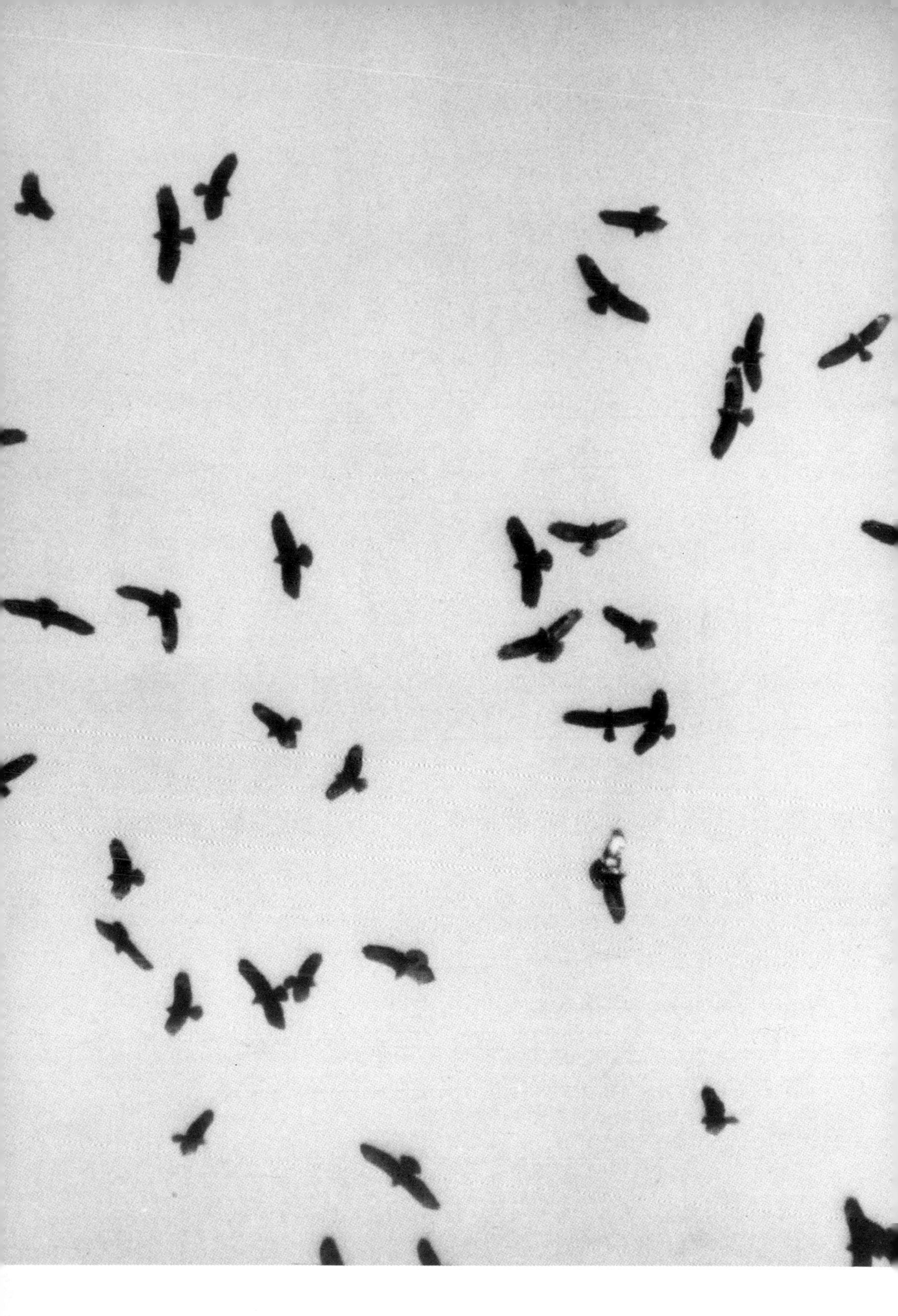

1 Buzzard *Buteo buteo*

Soaring flock of Buzzards on migration, Falsterbo, Sweden/5th October 1962 (Arthur Christiansen).

2 and 3 (*opposite*) **Buzzard** *Buteo buteo*

2 *Top*, pale rusty-yellowish form of the Steppe Buzzard *Buteo b. vulpinus*; *lower left*, juvenile grey-brown form of Steppe Buzzard, Israel/March (S. Christensen); *lower right*, adult grey-brown form of Steppe Buzzard, Israel/April (Claus F. Pedersen). Note in the adult the usually very broad conspicuous blackish band along trailing edge of wings on otherwise white primaries and whitish, slightly barred secondaries. The rusty-yellowish form is very like a Long-legged Buzzard *Buteo rufinus*.

3 *Upper*, immature typically plumaged bird with variably marked coverts and unevenly blotched brown body, heaviest markings being on the upper-breast, England/June (C. Pearson Douglas); *lower*, juvenile, a pale and lightly marked individual, typical of the northern European populations. Note the blackish moustachial streak, a common feature of light individuals which are distinguished from Rough-legged Buzzard *Buteo lagopus* by the many fine dark bars in tail (J. B. & S. Bottomley).

4 Buzzard *Buteo buteo*

Top, well marked adult bird soaring, Sweden (Jens Bruun); *above*, a pale juvenile, Falsterbo, Sweden/September (Arthur Christiansen); *lower left* and *right*, grey-brown Steppe Buzzards of *intermedius* type, the left bird showing typical pale orange slightly barred tail. Note pale whitish primaries, pale median wing-coverts and dark primary-coverts, the thick-set neck and body are well seen from directly below. Being juveniles they lack the clear-cut broad blackish band to trailing-edge of wings and tail, body is also longitudinally streaked. Turkey/September (M. J. Helps).

5 **Buzzard** *Buteo buteo*

Above, adult with Raven *Corvus corax* (John Marchington); *lower left*, juvenile, France/August (Pierre Petit); *lower right*, adult, Spain/March (S. Avellà).

All are typical birds of the race *B. b. buteo* from Western Europe.

6 Buzzard *Buteo buteo*

Adult Steppe Buzzards *B. b. vulpinus* from above showing pale base to primaries and variations in amount of pale at base of tail.
Note dark band on trailing-edge of wings and dark sub-terminal band on tail wider than other bars, characteristics of adults.

Israel/Spring (M. Müller and H. Wohlmuth).

7 Long-legged Buzzard *Buteo rufinus*

Above, juvenile without clear-cut blackish band to trailing-edge of wings and with washed out barring to flight-feathers; a rather pale bird resembling Rough-legged Buzzard *Buteo lagopus*, but tail practically unbarred. Pakistan (Eric Hosking); *centre* and *lower*, adult (Alfred Limbrunner). Note the dark belly, trailing-edge to otherwise white primaries and whitish dark barred secondaries, blackish carpal patches and very pale heads.

8 Long-legged Buzzard *Buteo rufinus*

Upper, adult blackish-brown phase India/February (Don Smith); this form which is very similar to dark Steppe Buzzard has not been recorded in Europe; *lower*, juvenile, Israel/Spring (M. Müller and H. Wohlmuth). Note moult into adult flight feathers has just started.

9 Long-legged Buzzard *Buteo rufinus*

Adult (unmarked tail and clear-cut blackish trailing-edge to wings).

Note faint moustachial and eye streak, often seen in this species (Ekko Smith). The Long-legged Buzzard is longer winged than Buzzard (including Steppe Buzzard) with often longer 'fingers'. Soars with wings lifted and pressed more forward than Buzzard, giving Golden Eagle *Aquila chrysaetos*-like appearance.

10 (*opposite*) and **11 Rough-legged Buzzard** *Buteo lagopus*

10 *Upper left* and *right*, two juveniles showing very pale, slightly barred whitish flight-feathers, whitish rather unmarked underwing-coverts, large blackish-brown belly patch, large blackish carpal patches and diffuse dark trailing edge to wings, Denmark (Jens Bruun); *lower*, adult low overhead, showing typically very dark throat, whitish U on breast, blotched underwing-coverts, rather barred flight-feathers and usually blackish well-marked trailing edge to wings. The narrow well-marked bars to tail diagnostic of adults (Emil Lütken and Carl Johan-Junge).

11 *Top*, juvenile, soaring, Falsterbo, Sweden/October (Jens Bruun); *lower*, juvenile hovering, Holland/December (Frits van Daalen). Note large whitish primary patches above, light median coverts, whitish uppertail-coverts and a single subterminal bar, diagnostic of juveniles.

12 Rough-legged Buzzard *Buteo lagopus*

All adults. *Upper*, Norway/July (P.-G. Bentz); *lower left* (note unusual absence of dark belly-patch), and *lower right*, Sweden (Jens Bruun).

The adult's characteristic tail bands and dark trailing-edge to flight-feathers can be seen in all pictures.

13 Honey Buzzard *Pernis apivorus*

Top, adult with juvenile Buzzard, France/August (Pierre Petit); *lower*, adult, Turkey/September (R. F. Porter).

Typical barred individuals. Note characters of adult: clear-cut tail-bands and broad blackish band along trailing-edge of wings, which otherwise have a much cleaner appearance than those of juvenile.

14 Honey Buzzard *Pernis apivorus*

Top, pale adult, Gibraltar/September (Clive Finlayson); *lower left*, barred adult, France/August (Pierre Petit); *centre right*, dark adult in typical gliding position, Turkey/September (R. F. Porter); *lower right*, barred adult, France (Pierre Petit), showing typical tail pattern clearly. Compare soaring and gliding birds, when long tail is striking. Note also slighty convex sides to tail which has rounded corners, and particularly small Cuckoo *Cuculus canorus*-like head and neck. Against the light, Honey Buzzards always show a large translucent patch confined to just the primaries.

15 Honey Buzzard *Pernis apivorus*

Juveniles, Sweden (Jens Bruun).

Note difference in thc spacing of tail bands compared to adult; also barring on secondaries and lack of broad dark trailing-edge to flight-feathers. The top bird shows translucent primary patch characteristic of Honey Buzzard when seen against light and which can assist with identification of birds seen at a distance.

16 Golden Eagle *Aquila chrysaetos*

Immature bird, Sweden/November (Aimon Niklasson).
Note the prominent white wing and tail patches, and soaring position with raised wings.

17 Golden Eagle *Aquila chrysaetos*

Upper, immature, Sweden/January (Karl-Erik Fridzén); *lower*, immature bird soaring, Sweden/November (Aimon Niklasson). Both photographs show the distinct white markings on the upper-wing and tail.

18 Golden Eagle *Aquila chrysaetos*

Upper left, juvenile, Sweden (Jens Bruun); and *upper right*, adult, Sweden/December (Arne Jensen); *lower*, immature mobbed by Goshawk *Accipiter gentilis*, Sweden/Winter (M. Müller and H. Wohlmuth).

Note adult's all dark plumage (though paler flight-feathers), large tail, and wings raised and pressed

19 Spotted Eagle *Aquila clanga*

Upper, same immature in two different flight situations, Sweden/October (Jens Bruun); *centre* and *lower*, immature, S. France/Winter (Charles A. Vaucher).

Note underwing-coverts below are darker than flight-feathers (the opposite in Lesser Spotted Eagle) and white crescents at base of primaries. Apart from white tips to coverts and flight-feathers and white shafts to primaries the upperwing is all dark, with no contrast between coverts and flight-feathers, another important point of distinction from Lesser Spotted (see Plate 22).

20 Spotted Eagle *Aquila clanga*

Above, Sweden (Arne Schmitz); *lower left*, Sweden/October (Søren Breiting); *lower right*, immature, Oman/December (John Warr).

The comparatively broad wings and fairly short, sometimes wedge-shaped tail in worn plumage, give shape not unlike miniature White-tailed Eagle *Haliaeetus albicilla*, but head and neck protrudes much less. Underwing-coverts normally clearly darker than flight-feathers.

21 Spotted Eagle *Aquila clanga*

Upper, sub-adult, India/Winter (Ekko Smith), white crescents show clearly around carpal patch and note contrast between dark coverts and flight-feathers; *lower*, juveniles, India/January (R. F. Porter); note in addition to points mentioned above, the translucent line on trailing-edges of wings and tail.

22 Lesser Spotted Eagle *Aquila pomarina*

Above, distinctive immature plumage, Lebanon/September, and *below left*, adult bird, Lebanon/September, *below right*, immature in transitional plumage, Israel/April (Claus F. Pedersen).

Note contrast between paler coverts and darker flight-feathers, above and below, and compare with Spotted Eagle (Plates 19–21).

23 Lesser Spotted Eagle *Aquila pomarina*

Upper, adult, Germany/August (B.-U. Meyburg). Note contrast between medium brown coverts and blackish flight-feathers, small pale 'flash' at base of primaries and light uppertail-coverts; typical soaring position with bowed wings; *lower*, Lebanon/September, showing typical gliding attitude (Claus F. Pedersen).

24 Lesser Spotted Eagle *Aquila pomarina*

Upper and *lower*, immature soaring, Turkey/September (Wolfgang Wagner).

25 Imperial Eagle *Aquila heliaca*

Upper, immature (1 year old) with Buzzard *Buteo buteo*, Sweden/August (P. G. Bentz); *lower*, immature, Saudi Arabia/Winter (N. R. Phillips).

Note plumage of juvenile and young immatures with pale underbody and coverts contrasting with blackish flight and tail feathers; streaked breast demarcated from pale lower body and palish inner primaries usually distinctive.

26 Imperial Eagle *Aquila heliaca*

Top, adults of nominate race, Israel/December (R. F. Porter), note contrasting dark coverts and body, particularly noticeable in strong sunlight; *lower left*, bird in juvenile plumage, Turkey/September (M. J. Helps); *lower right*, immature in transitional plumage, Israel/April (Clause F. Pedersen).

Note long wings, longish tail (but less than Golden) and well protruding head. The plumage of the juvenile bird is distinct, but the streaks across the breast are commonly extended to whole belly. Note the pied appearance of the immature bird and the still pale inner-primaries, vent and undertail-coverts; yellowish head starts to appear.

27 Imperial Eagle *Aquila heliaca*

Above and *lower left*, juvenile/immature from above, India/April (T. Shiota). Note similarity to juvenile Steppe Eagle *A. nipalensis* (Plate 29), but Imperial has streaked back and wing-coverts; *lower right*, immature in early transitional plumage (third year); in this plumage dark blotched forebody contrasts with pale unstreaked hindbody; dark feathers start to appear in wing-coverts. Israel/Spring (M. Müller and H. Wohlmuth).

28 Imperial Eagle *Aquila heliaca*

Adult of Spanish race *A. a. adalberti*, Spain (Charles A. Vaucher).

29 Steppe Eagle *Aquila nipalensis*

Top left, juvenile, Kenya/February (S. Christensen and Bent Pors Nielson); *top right*, juvenile, Israel/Spring (M. Müller and H. Wohlmuth); *lower*, juvenile/immature, India/Winter (Ekko Smith).

White band along middle and rear edge of wings and tail characteristic of juvenile and from above a whitish patch at base of primaries merging with trailing-edge, otherwise upper parts similar to Imperial but un-streaked (see Plate 27).

30 Steppe Eagle *Aquila nipalensis*

Top, immature ($1\frac{1}{2}$ year old), Tanzania/January (B.-U. Meyburg); *lower left* and *right*, immatures, Israel/Spring (M. Müller and H. Wohlmuth). Band down centre of underwing becomes ragged and bands on upperwing less obvious. Barring on flight-feathers just discernible.

31 Steppe Eagle *Aquila nipalensis*

Top, immature of pale variant in transitional plumage, still showing white primary coverts (Claus F. Pedersen); *centre*, adult and *lower*, sub-adult moulting into adult plumage, Israel/Spring (H. Wohlmuth)

Note in adult/sub-adult heavily barred flight and tail feathers, more so than in any other Aquila, bordered by dark band on trailing-edge typical of most adults. Pale throat patch in *top* and *centre* birds, bordered by dark line below yellow gape flange is again typical in most Steppes. *Centre* and *lower* birds each show darker body, carpal area and primary tips, a characteristic of most adults and sub-adults.

32 Steppe Eagle *Aquila nipalensis*

Top, adult; *lower left*, sub-adult (at paler end of range of plumage); *lower centre* and *right*, adult; all photos Israel/Spring (M. Müller and H. Wohlmuth).

For main features of adult see caption to Plate 31; note especially barred flight and tail feathers both above and below and that white edgings to feathers of greater coverts linger on into adult plumage in many birds.

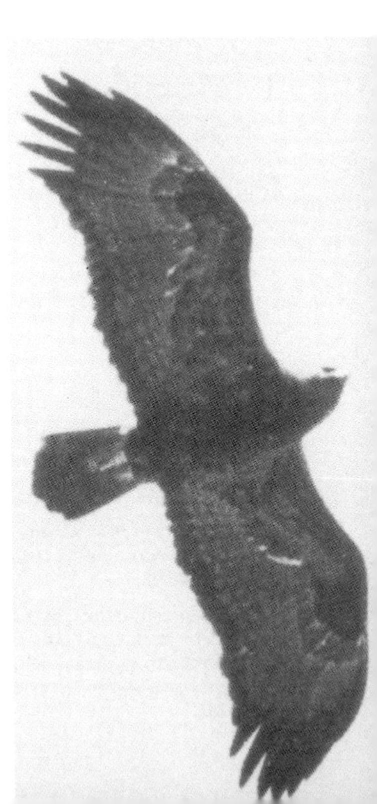

33 White-tailed Eagle *Haliaeetus albicilla*

Upper left and *right*, adults, Norway/May and June (Eric Hosking, Gunner Lid); *lower*, juvenile, Fair Isle (R. J. Tulloch). Note huge wings, heavy bill and long neck with short all white tail in adult and slightly longer tail of juvenile with whitish centres to dark feathers.

34 White-tailed Eagle *Haliaeetus albicilla*

Upper, immature, Denmark/May; this 11 month old bird has moulted into dark and pale mottled-plumage of immature; note pale axillary patch and bar on median coverts (M. Müller and H. Wohlmuth); *centre*, immature, Sweden/Winter; note largely pale bill and dark tail when closed (M. Müller and H. Wohlmuth); *lower*, immature in transitional plumage, Sweden/Winter (Jens Bruun); note white in spread tail feathers, variegated pattern on back and median/lesser coverts and pale inner greater-coverts and inner secondaries. Massive bill and long neck characteristic of all ages.

35 Booted Eagle *Hieraaetus pennatus*

Light phase, France/August (Pierre Petit); *inset*, adult above, Spain/March (S. Avellà).

36 Booted Eagle *Hieraaetus pennatus*

Upper left, dark phase, France/August (Pierre Petit); *upper right*, dark phase, Spain/November (Riccardo Nardi); *lower*, light phase, Spain/March (S. Avellà).

Note structure not unlike Buzzard but with 5 'fingers' (only 4 in *Buteo*), more parrallel-edged wings and square-cut, sharp-cornered tail. Light phase should present few problems but dark phase can resemble Black Kite *Milvus migrans* and Marsh Harrier *Circus aeruginosus*, but combination of rounded tail, soaring and gliding on flattish wings, transluscent trailing wedge at junction of primaries and secondaries and transluscent trailing-edge to wing and tail, distinguish dark phase Booted from these two species. If seen close in sunlight, some dark phase Booteds can show warm brown underbody and coverts with blackish greater coverts standing out as a dark band (see *upper right*).

37 Bonelli's Eagle *Hieraaetus fasciatus*

Above, adult, Spain (Charles A. Vaucher); *below*, adult, Spain/May (Pierre Petit).

Note the broad wings, long tail and well protruding head, in general not unlike a thick-set Honey Buzzard *Pernis apivorus*.

38 Bonelli's Eagle *Hieraaetus fasciatus*

Upper, adult, Tunisia/March (Riccardo Nardi); *lower left*, juvenile/immature, Israel/Spring (M. Müller and H. Wohlmuth); *lower right*, adult, Spain/May (I. R. Hornsby).

Note the pale body of adult contrasting with largely dark underwings (but note pattern) and dark terminal band to tail. In juvenile the dark band around carpal area and along edge of coverts is characteristic.

39 Short-toed Eagle *Circaetus gallicus*

Upper, France/May (Pierre Petit); *lower*, Israel/spring (M. Müller and H. Wohlmuth). Very pale plumage below with dark patterning and bands on tail both above and below. Note primary moult in upper bird; there are normally 5 free 'fingers'.

40 and **41** (*opposite*) **Short-toed Eagle** *Circaetus gallicus*

40 *Upper*, male above, and female, France/April (Pierre Petit); *lower*, pale individual, Turkey/September (Wolfgang Wagner).

41 *Upper*, hovering, Jordan/April (Eric Hosking); *lower left*, migrant, Turkey/September (M. J. Helps); *lower right*, migrant, Lebanon (Claus F. Pedersen). Except for bird shown lower right, all these are dark-headed birds with heavily patterned under-wings.

42 (*opposite*) and **43 Osprey** *Pandion haliaetus*

42 *Upper,* migrant, France/August (Pierre Petit); *lower* (Runar Söderberg).

43 *Above*, immature (first year) on migration, France/September (Pierre Petit); *lower*, adult, S. Sweden/July (I. R. Hornsby).

Note the pure white underparts, characteristic head markings, dark band along middle of wing, and black carpal patch. When seen against the light, the spread tail becomes whitish and translucent. Note speckling on upperparts of immature.

44 Marsh Harrier *Circus aeruginosus*

Above left, juvenile, Pakistan/December (Eric Hosking); *above right*, adult male, Denmark (Gert Østerbye), *below*, immature female, Sweden/May (Axel von Arbin).

Note wings in shallow V when gliding; also in juvenile (and female) dark coverts contrasting with slightly paler flight-feathers – a useful feature to separate Marsh Harrier from dark phase Booted Eagle when observed high over-head.

45 Marsh Harrier *Circus aeruginosus*

Above left, adult male (Claus F. Pedersen), *above right*, immature male, England/April (Eric Hosking); *below*, immature female, Sweden/May (Axel von Arbin).

46 (*opposite*) and **47 Hen Harrier** *Circus cyaneus*

46 Adult female, France/August (Pierre Petit).

47 *Upper*, male, France/June (Pierre Petit); *lower*, female, Denmark (Arthur Christiansen).

In male, note white underparts, grey head, black primaries and dark trailing edge to translucent secondaries. Note female's long and fairly broad wings, somewhat owl-like head bordered by 'ruff' of streaks extending onto underparts, and bars on primaries, secondaries and tail.

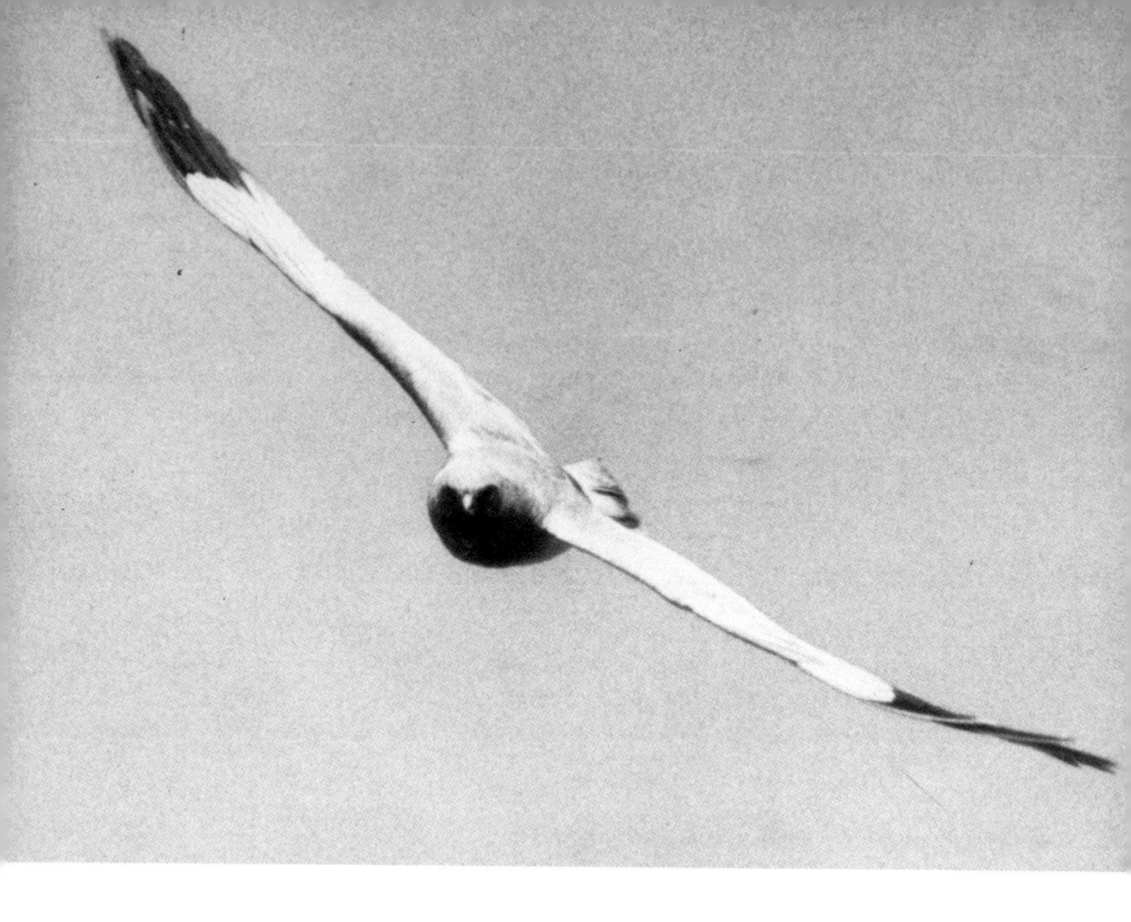

48 Hen Harrier *Circus cyaneus*

Above, male attacking photographer near nest, France/June (Pierre Petit); *below*, immature, Denmark (Arthur Christiansen).

49 Montagu's Harrier *Circus pygargus*

Above, male, Sweden/May (Karl-Erik Fridzén), showing mid-grey upperparts, black primaries, narrow black bar bordering coverts, wings in shallow V; *below*, male, Denmark (Arthur Christiansen), with dark grey head, rusty streaks on white lower breast, black primaries, black bars on secondaries and patchy flecking on white coverts.

50 Montagu's Harrier *Circus pygargus*

Upper, male, Israel/Spring (M. Müller and H. Wohlmuth); note slim wings and long tail; *lower*, female, England/June (Eric Hosking); note well-spaced dark bars on secondaries, compare this and head pattern to that of juvenile or female Pallid Harrier *Circus macrourus*.

51 Montagu's Harrier *Circus pygargus*

Upper, male, Spain/April (R. C. Homes); *lower*, female, India/April (T. Shiota).

Note in lower bird and female on Plate 50 dark cheek-patch, whitish 'eye-mask' and absence of pale collar (compare head pattern to that of juvenile and female Pallid Harrier).

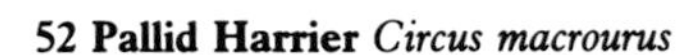

52 Pallid Harrier *Circus macrourus*

Upper left and *right*, male, Sweden/June (Erik Engqvist); note slim build, all white below except for narrow black wedge on primaries; *lower*, juvenile, January (Rudi Jelinek); note pale chestnut-buff underparts and underwing-coverts, barred primaries and tail and contrasting head pattern (also characteristic of female) with pale collar accentuated by narrow border of streaks across throat (cf Montagu's Harrier).

53 Pallid Harrier *Circus macrourus*
Male, Roumania (Arthur Christiansen).

54 and **55** (*opposite*) **Red Kite** *Milvus milvus*

54 Juvenile, Sweden (Frank Bergström).

55 *Top left*, adult, Sweden (Jens Bruun); *top right*, adult, mobbed by Carrion Crow *Corvus corone corone* (Helmut Blesch); *centre*, adult, Spain/April (B.-U. Meyburg); *bottom*, juvenile, Sweden (Frank Bergström).

Note conspicuous white patches at base of primaries, deeply forked tail, manoeuvred constantly, and arched wings when gliding. The juvenile has paler body, well streaked breast, pale undertail-coverts not contrasting to pale undertail and darker bar along tip of of tail. Adult has darker body, more contrasting to pale head and undertail.

56 (*opposite*) and **57 Black Kite** *Milvus migrans*

56 *Top*, adult (Eric Hosking); *below*, feeding over Bosphorus, with Herring Gulls *Larus argentatus* (Ian Willis).

57 *Above*, juvenile (Eric Hosking); *lower left*, taking offal from water (J. F. Ormond); *right*, seen from above (Claus F. Pedersen).

Compare the straight rear edge of Black Kite's spread tail with Red Kite's ever-present fork. Note soaring silhouette with small head, triangular tail, and wings slightly forward; pale diagonal band on greater coverts when seen from above. Note juvenile's tail shape, pale streaked underbody, and lighter markings on underwing.

58 (*opposite*) and **59 Black Vulture** *Aegypius monachus*

58 *Above*, Spain/March (B.-U. Meyburg); *below*, Majorca (S. Avellà).

59 *Above*, in silhouette (Stefan Danko); *below*, soaring bird seen head-on, Spain/March (B-U. Meyburg).

Note all-black plumage with wing coverts and body darker below than flight-feathers, and conspicuous pale feet. Wings are broad, without bulging rear edges of Griffon Vulture *Gyps fulvus*; when soaring, wings are held flat, not raised as in Griffon. Note also pointed flight-feathers (cf Griffon Vulture).

60 (*opposite*) and **61 Griffon Vulture** *Gyps fulvus*

60 Griffon Vultures, soaring, Spain/May (Charles A. Vaucher).

Note marked curves to rear edges of wings, formed by bulging secondaries; also the whitish head and ruff, and marked contrast between sandy body and wing coverts and dark flight feathers and tail; also, soaring on raised wings.

61 Griffon Vulture, Lammergeier *Gypaetus barbatus* and Egyptian Vulture *Neophron percnopterus* soaring together, Spain/May (Charles A. Vaucher).

62 (*opposite*) and **63 Griffon Vulture** *Gyps fulvus*

62 *Upper*, juvenile, Israel/April (Claus F. Pedersen); *lower*, adult mobbed by Bonelli's Eagle *Hieraaetus fasciatus*, Spain/May (Charles A. Vaucher).

63 *Above*, view from overhead, West Pyrenees/April (P. Van Groenendael and W. Suetens); *below*, from overhead, Spain/May (Charles A. Vaucher).

Note pale head, whitish ruff, sandy-grey back and wing coverts, contrasting with black flight feathers and tail.

64 and **65** (*opposite*) **Lammergeier** *Gypaetus barbatus*

64 *Above*, adult, Israel/March (Claus F. Pedersen); *below*, adult, Kenya (Rudi Jelinek).

65 *Upper*, adult, Kenya (Rudi Jelinek); *lower*, sub-adult, Mongolia/June (B.-U. Meyburg).

Note long wings, tapering at hands, and long, ample, diamond-shaped tail; yellowish-orange head (in adult) and underparts contrast with black eye-stripe and moustaches, slatey-black wings and tail, and black underwing coverts.

66 (*opposite*) and **67 Egyptian Vulture** *Neophron percnopterus*

66 *Upper left*, immature, Spain/April (B.-U. Meyburg); *upper right*, adult, Israel/April (Claus F. Pedersen); *lower left*, adult, Spain/May (Charles A. Vaucher); *lower right*, immature (Charles A. Vaucher).

67 *Top*, adult, Israel/April (Claus F. Pedersen); *lower*, two immature birds, Spain/May (Charles A. Vaucher).

Note characteristic black and white plumage of adults, with dirty yellowish discolouration on the head and upper breast, yellow face, and wedge-shaped tail which often appears translucent in strong sunlight. Note, also, dark brown smudgy plumage of immature.

68 (*opposite*) and **69 Gyrfalcon** *Falco rusticolus*

68 *Upper left* and *lower*, immature, Sweden/March (Karl-Erik Fridzén); *upper right*, juvenile, Sweden (J.-M. Breider).

69 *Above*, Iceland (Paul Géroudet); *below*, in hot pursuit of drake Mallard *Anas platyrhynchos*, Sweden/March (Karl-Erik Fridzén).

The largest falcon—compare with Mallard—with heavy, powerful structure and broad based wings. Grey phase shown here. Dark ear-coverts merging into fairly pronounced moustachial streak can suggest Peregrine *F. peregrinus*, but this effect is lost in adults.

70 Saker *Falco cherrug*

Above left, adult, Czechoslovakia/May (Ján Svehlik); *above right*, Turkey (A. R. Kitson); *lower left* and *right*, probably an escaped falconer's bird, Shetland/Autumn (Bobby Tulloch).
Large and rather heavy, with long fairly broad-based but pointed wings, the dark underwing coverts contrasting with flight feathers, especially in juveniles.

71 Saker *Falco cherrug*

Above and *below*, Czechoslovakia (Kovář Karel).

This species has a pale, sometimes nearly white, head with streaks on crown and a narrow rather indistinct moustachial streak, variable longitudinal spotting on underparts and underwing coverts (lighter on forewing), and quite conspicuously barred tail.

72 Lanner *Falco biarmicus*

Above, Tanzania/April (Rudi Jelinek); *right*, captive bird, England/April (Robin Williams); *lower left and right*, Rhodesia (Peter Steyn).

Note the pale rufous crown of lower birds and quite unmarked underparts of upper are not typical in *feldeggii* the race of Mediterranean Europe, and birds pictured here are the rather different nominate African race *F. b. biarmicus*.

73 Peregrine *Falco peregrinus*

Left (Karl-Erik Fridzén); *right* (both) (Eric Hosking).

More compact than Lanner, with broader-based pointed wings, conspicuous head pattern (very dark crown and moustachial patch contrasting with white cheeks, chin and throat); grey-barred underbody and densely barred and spotted underwing showing little contrast between flight-feathers and coverts; juveniles have buffish underparts with brown streaking. Note pointed wing-tip in soaring bird and compare with rounded wing-tip in soaring Gyr, Saker and Lanner.

74 Eleonora's Falcon *Falco eleonorae*

Top left, juvenile, Cyprus/October, and *centre left*, light phase adult, Cyprus/September (P. R. Flint); *lower left*, light phase female, Crete/Autumn (Philip Round); *top right*, adult pale phase in silhouette, Majorca (Arthur Christiansen); *centre right*, dark phase adult from above; *lower right*, juvenile, Cyprus/October (P. R. Flint). Noticeably larger than Hobby with relatively long, rather narrow wings and long tail. From below, pale flight-feathers bordered by broad dark trailing-edge and dark-spotted wing-coverts, diagnostic of juveniles. Adults have very dark underwing-coverts, paler flight-feathers, often darkening towards rear edge of wings, sometimes producing a diffuse pale "bar" to centre of midwing and paler based hand. The flight may be slow and relaxed, or swift and agile.

75 Eleonora's Falcon *Falcon eleonorae*

Top left and *right*, juveniles, Majorca/October (S. Avellà); *lower*, light phase adult, Majorca (J-M. Breider).

76 Hobby *Falco subbuteo*

Top, Sweden/Spring (Aimon Niklasson); *lower*, catching insects, Turkey/September (R. F. Porter).

77 Hobby *Falco subbuteo*

Top left, Turkey/September (R. F. Porter); *lower left*, Sweden/Spring (Aimon Niklasson); *right*, Slovakia/August (L. Šimák).

Medium-small and slender, with relatively short tail and narrow very pointed wings, giving silhouette not unlike large Swift *Apus apus*. Adults have clearly defined, unstreaked, rich rufous-red patch on thighs and under-tail coverts, and at all ages the marked moustachial streaks, rather plain dark upperparts, unbarred uppertail, heavily streaked underbody and uniformly greyish underwings (at distance), help identification.

78 Red-footed Falcon *Falco vespertinus*

Top, adult female (Charles A. Vaucher) and *inset*, adult female, Cyprus/September (P. R. Flint); *below left*, juvenile, Denmark/August (Knud Pedersen); *below right*, juvenile, Sweden/September (L. Eccles).

Note the diagnostic unstreaked rufous-yellow underparts of adult female. Juvenile may superficially resemble juvenile Hobby but note paler ground colour to flight-feathers, palish or yellowish-brown forehead and crown with darker streaks and barred upper tail. Some adult females (and juveniles) show rather conspicuous dark trailing-edge to underwings (see *top* birds).

79 Red-footed Falcon *Falco vespertinus*

Top left, sub-adult male with variegated pattern of moulting flight-feathers, Bosphorus/September (Henrik Korning); *top right*, adult male (Kovář Karel) and, *inset*, sub-adult male, same stage as left bird, Bosphorus/September (Steen Søgård); *below*, juvenile, Denmark/August (Knud Pedersen); note barred uppertail (not in Hobby).

Adult males are variable sooty grey beneath, except for dark red flanks and undertail-coverts; flight feathers all dark grey, slightly contrasting with darker grey coverts. The variegated pattern birds above are third autumn males, moulting into adult plumage. Note kestrel-like contrast above of juvenile and pale sides of neck joining pale nape.

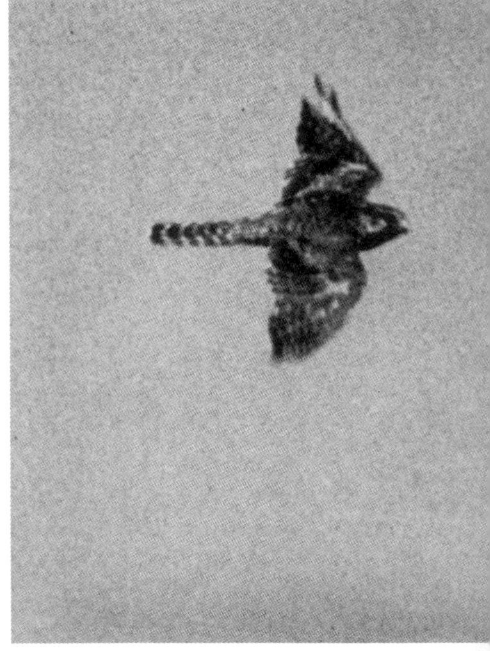

80 Merlin *Falco columbarius*

Top (Arthur Christiansen); *lower left*, Norway (J.-M. Breider); *lower right*, Shetland (Bobby Tulloch).

A compact little falcon with small broad-based wings. Male is smallest European raptor, but larger female may overlap in size with other small falcons.

81 Merlin *Falco columbarius*

Top left, Sweden (Arne Jensen); *top right*, male, Wales/Summer (J. Lawton Roberts); *lower left*, male (J. Lawton Roberts) and, *lower right*, juvenile, Scotland (Bobby Tulloch).

Note the comparatively short wings and longish tail.

82 Kestrel *Falco tinnunculus*

Top, hovering female (J. B. and S. Bottomley); *below*, adult male (Derek Washington).

One of the commonest and more widely distributed of European raptors, habitually hovering with rapid shallow wing-beats, and fanned tail.

83 Kestrel *Falco tinnunculus*

Top left, female, Turkey/September (Wolfgang Wagner); *top right*, female, England (Robin Williams); *lower left*, adult male, Libya (Arthur Christiansen); *right*, adult male, England (Barry Angell).

84 Lesser Kestrel *Falco naumanni*

Top, hovering male, Spain/June (Paul Géroudet); *below*, juvenile, Gibraltar/July (Clive Finlayson).

Lesser Kestrel is slightly slimmer than Kestrel *Falco tinnunculus*, and frequently has wedge-shaped tail. Note paleness of juvenile's underwing (compare with female Kestrel).

85 Lesser Kestrel *Falco naumanni*

Top left, male, Spain/May (Pierre Petit); *top right*, female, Spain/Summer (Jens Bruun); *below left*, male (J. F. Ormond); *right*, male carrying insect in bill, Turkey (M. J. Helps).

Note male's creamy-buff body, white underwings lightly spotted, and black wing-tips, unspotted chestnut back and upper-wing coverts, and contrasting blue-grey greater coverts. Kestrel's moustachial stripes are lacking. Wedge-shaped tail can be seen clearly in two of the birds shown. The upperparts of female are very similar to those of Kestrel.

86 and (*opposite*) **87 Black-winged Kite** *Elanus caeruleus*

86 ***Top*** **and** ***below*****, immature hovering, Kenya/May/March** (J. F. Reynolds).

87 *Top*, Kenya (Arthur Christiansen); *below*, Portugal (D. Wooldridge).

Longer-winged and somewhat larger than Kestrel *Falco tinnunculus*, with noticeably protruding owl-like head and fairly short tail; hovers persistently; soars with wings held forward and arms raised.

88 Black-winged Kite *Elanus caeruleus*

Top left, immature, Kenya/May (J. F. Reynolds); *top right*, adult, Gambia/January (Bent Pors Nielsen); *below*, adult, Rhodesia (Peter Steyn).

Plumage basically grey and white, with black upperwing patches and black under-primaries.

89 Goshawk *Accipiter gentilis*
Adult, Sweden (Arne Schmitz).

90 Goshawk *Accipiter gentilis*

Above left, immature, Sweden (J.-M. Breider); *above right*, adult Sweden (Arne Schmitz); *below*, immature, Czechoslovakia (Kovář Karel). Great size difference between the sexes; the smallest males may be a little larger than female Sparrowhawk *Accipiter nisus*. Immature heavily chequered below, adult barred; latter's dark crown and ear coverts may give it a hooded appearance. The male Goshawk has deeper chest, more protruding head and comparatively shorter, more rounded tail than female Sparrowhawk.

91 Sparrowhawk *Accipiter nisus*

Top and *bottom*, Falsterbo, Sweden/October (Jens Bruun).

92 Sparrowhawk *Accipiter nisus*
Top, Skagen/May (Knud Pedersen), *below* (Gert Østerbye).

93 Sparrowhawk *Accipiter nisus*

Top, mobbed by Starlings *Sturnus vulgaris* (Karl-Erik Fridzén); *below*, Sweden (Arne Schmitz).

94 Sparrowhawk *Accipiter nisus*

Female, hunting (Arthur Christiansen).

This and the Kestrel *Falco tinnunculus* are the most numerous and widespread of small European raptors. The much larger female may approach small male Goshawk *Accipiter gentilis* in size, though more compact and with square-ended tail. In hunting flight employs series of several rapid wing-beats and stiff, short glides. Note 6 emarginated primaries compared to 5 in Levant Sparrowhawk *A. brevipes*.

95 Levant Sparrowhawk *Accipiter brevipes*

Top left, immature, Turkey/September (Peder Weibull); *top right*, immature, Turkey/September (Claus F. Pedersen); *lower left*, adult male in partial silhouette, Turkey/September; *lower right*, adult male, Turkey/September (M. J. Helps).

Similar in outline to Sparrowhawk *Accipiter nisus*, but silhouette more falcon-like; adult male shows unmarked white below, suffused pinkish, and black wing-tips, but identification of female needs more care (see text). Immature has lines of drop-like spots on pure whitish underbody, unlike ragged barring of young Sparrowhawk. Frequently seen in large flocks on migration.

96 Levant Sparrowhawk *Accipiter brevipes*

Top, adult female, Turkey/September (M. J. Helps); *below*, adult male, Turkey/September (R. F. Porter).

Not the wing formula which can be clearly seen in upper bird; 5 emarginated primaries (compared with 6 in Sparrowhawk) helping to produce more falcon-like silhouette. The plumage of the male with pink-tinged white underparts and black wing-tips is unmistakeable.